Thinking Race

Thinking Race

Social Myths and Biological Realities

Richard A. Goldsby and
Mary Catherine Bateson

ROWMAN & LITTLEFIELD
Lanham • Boulder • New York • London

Published by Rowman & Littlefield
An imprint of The Rowman & Littlefield Publishing Group, Inc.
4501 Forbes Boulevard, Suite 200, Lanham, Maryland 20706
https://rowman.com

6 Tinworth Street, London SE11 5AL, United Kingdom

British Library Cataloguing in Publication Information Available

Library of Congress Cataloging-in-Publication Data

Includes bibliographic references and index.
ISBN 978-1-5381-0501-6 (cloth : alk. paper)
ISBN 978-1-5381-0502-3 (Electronic)

∞ ™ The paper used in this publication meets the minimum requirements of American
National Standard for Information Sciences Permanence of Paper for Printed Library
Materials, ANSI/NISO Z39.48-1992.

This book is dedicated to the memory of three young men: James Chaney, Andrew Goodman, and Michael Schwerner, one of them Black and the other two White, who were martyred near Philadelphia, Mississippi, on the night of June 21, 1964, for their efforts in the struggle for equality for all Americans.

Contents

Preface

If you live in the United States, you and most of your friends probably assume membership in some race or another. After all, race is a pervasive feature of the American Experience. It was there in the early 1600s when the curtain was raised on the drama that became the United States, and it offers no promise of exiting the stage anytime soon.

But what is race? Is it something akin to being Republican or Democrat, labels with important social and cultural distinctions but having no basis in DNA? If this is the case, race is solely cultural, and biology and medicine have nothing to contribute to a conversation about it. On the other hand, if race is something mostly determined by DNA, and therefore biological, social construction will have little role in determining its nature. So is race social construct or biology?

We show here that like so many things, it is not all one or the other. Our discussion of the medical implications of race makes the connection between race and biology firm and clear. As we wonder about the origin of race and some of the characteristics associated with this or that race, both culture and biology will prove informative. However, when we explore the approaches to managing and perhaps even solving some of the societal problems that have grown out of race, we find biology largely irrelevant, offering little, if any, help.

We hope your journey through this little book allows you to understand race as yet another manifestation of the enormous cultural and much smaller biological variety found in our species. Have an informative and surprising trip.

Acknowledgments

We are grateful to the many friends and colleagues who have generously given advice and offered opinions that have helped us think about race and about this book presenting our thoughts on race. We thank Johnnetta Cole, Joan Godsey, Carl Krause, Thomas Kindt, Joanna Wilbur, Charles Osborne, Barbara Rell, Robert Shoenberg, Susan Houchin, Leodis Davis, June Wilson Davis, Joan Weiss, Meng Zhang, and William Zimmerman for their thoughtful reading of various chapters and sections of our book. The detailed edits, advice, and suggestions provided by David Ratner, Amy Weinberg, Lydia Villa-Komaroff, Gerald Fink, and Richard Godsey were very helpful. We thank Meng Zhang for helpful conversations on population differences in drug efficacy and toxicity. We are deeply grateful to our respective spouses, especially to Barbara Osborne who generously, patiently, and thoughtfully read successive drafts of our manuscript offering advice and encouragement during difficult stages of the project, always providing useful suggestions, and to J. B. Kassarjian, who urged us toward thematic coherence. Robert Weinberg's gift of many reprints and commentaries examining biological diversity in a variety of populations and ethnic groups was very helpful in the early formative stages of our writing. We thank George Greenstein for directing us to the work of Martha Sandweiss on racial passing. We are indebted and grateful to Robert Wedgeworth for his critical help in placing our manuscript in the capable hands of Rowman & Littlefield. We are grateful for the material assistance of Amherst College's superb Beneski Museum of Natural History and to the generous and resourceful advice and help of Alfred Venne, museum educator, and Diane Hutton, administrative coordinator for the Amherst College Department of Geology. We also thank Jiayu Liu, the skilled and energetic Amherst College photographer.

But our debt to Amherst College goes further back. We met at Amherst College when one of us, Mary Catherine Bateson, was dean of the faculty, and recruited the other, Richard Goldsby, to a distinguished chair in the Biology Department, where he served as Amanda and Lisa Cross Professor, and where our friendship and collaboration (this is our second joint book) developed and came to include our respective families. Although we do not point to specific institutions, Amherst resembles many other fine academic institutions in having a history of striving to address inequities related to race.

Because we currently live in separate states (Massachusetts and New Hampshire), this collaboration has involved a lot of commuting and telephoning, as well as lunches and dinners rich with conversation and debate. Each of us is grateful to the other for what we have learned from each other and for our shared commitment to equality of opportunity for Americans who have traveled out of Africa over many millennia, both willingly and as captives, to take root and contribute richly to the resilience and creativity of this human community. We are especially grateful to a restaurant in Peterborough, New Hampshire, called Kodetsu, where two Americans—one black, one white, and both devoted to sushi and sashimi—have sorted out our ideas. In spite of the ugly history, we celebrate the diversity with which our country is blessed.

Chapter One

Generations of Migration

²Abraham begat Isaac; and Isaac begat Jacob; and Jacob begat Judas and his brethren;
³And Judas begat Phares and Zara of Thamar; and Phares begat Esrom; and Esrom begat Aram;
⁴And Aram begat Aminadab; and Aminadab begat Naasson; and Naasson begat Salmon;
⁵And Salmon begat Booz of Rachab; and Booz begat Obed of Ruth; and Obed begat Jesse;
⁶And Jesse begat David the king; and David the king begat Solomon of her that had been the wife of Urias; ⁷And Solomon begat Roboam; and Roboam begat Abia . . . And so on.
—Matthew 1:2–7 of the King James Bible

Whoever we think we are, if we trace back far enough, the family tree is rooted in Africa. Analysis of DNA sampled across the broad diversity of human populations shows a common African ancestry for modern humans, the term we will use interchangeably with *Homo sapiens*, our species of human. Wherever our current homeland, we all share a motherland in Africa. We know that somewhere on that vast continent the apes that are modern humans arose and humanity began its climb to dominance. Following our branch back down the family tree we find that five to seven million years ago, our lineage shared a common ancestor with chimpanzees, the smart, agile, and strong little apes Jane Goodall has told us so many fascinating things about. These are our closest living relatives. As different as these two creatures seem to be, comparisons of the DNA sequences of chimpanzees and humans reveal that almost 96 percent of our DNA sequence is identical to theirs. On average, almost ninety-six out of every hundred letters in the DNA sequence is the same in humans and chimps. This provides an ironclad chemical confirmation of our evolutionary similarity to this primate relative

1

with whom we share the planet. It also shows that even small differences can make a difference; in this case, a very big one. The 4 percent difference that separates us from them results in profound differences in outward appearance, the size and architecture of the brain, analytical ability, tool-making capacity, and even lifespan. We speak, they don't. Jane Goodall writes about chimps, they don't write about her.

A word or two about how we determine when things happened in prehistoric times. First of all, dates and time intervals are estimates, in most instances highly approximate ones. You will see many numbers with a lot of zeros. Though estimates, they are more than just surmise, guess, or folklore. They are based on data acquired using a variety of approaches, some of them highly advanced technologies of geology, archaeology, isotopic dating, and DNA sequencing. As is true of most approximations, we can expect that as more data is accumulated and new findings are made, our current estimates will be refined upward or downward. In some cases, the revisions will be dramatic. Between 2016 and 2017, new fossil finds changed the site at which the earliest remains of *H. sapiens* had been discovered from Ethiopia to Morocco and pushed back the time modern humans appeared on the world stage from around 200,000 years ago to around 300,000, a surprising and impressive 100,000 additional years. As powerful as the advanced scientific methods deployed to examine fossil finds are, they provide estimates, only accurate to the nearest 10,000 or 20,000 years, not exact dates. Modern science, with its approximations, equivocations, and acknowledged ignorance, must remain in humbled awe of James Usher, Archbishop of Ireland. Using only his Bible and a calendar, during the seventeenth century, this confident prelate calculated the date of God's completion of the creation of the world to nightfall on October 22, 4004 BCE. It was a Saturday.

We certainly began as small populations, weak and scattered, with neither tribes nor nations. There were just bands, bound by kinship and proximity, perhaps similar to the troupes of contemporary chimpanzees. It was a tenuous and vulnerable existence that could have been wiped out by something as dramatic as predation by carnivores or by events as random as an accident or tragic as starvation. Any of these calamities and others we could imagine might have terminated or reduced the fledgling population of humans to unsustainable numbers. It is likely that these early bands would have occupied small territories, such that severe weather events—violent storms, devastating floods, prolonged drought—might have been disasters they couldn't survive.

Even love, or more likely just sex, across species differences, a matter we'll revisit later in the chapter, could have been a gentler but still lineage-ending calamity for emerging modern humans, changing uniquely human gene pools to something else. For a time, we surely coexisted with the earlier versions of humans from whom we arose. In the early stages of our emergence, carnal connections with our closely related forerunners who were then contemporaries would have carried the very real possibility of our founding ancestral population being reabsorbed by related populations with whom they would have been interfertile. However, despite the many roads leading to extinction, our fate was the narrow path of survival. And here we are, a species likely to number ten billion by 2050 and a major, if not always a responsible, ecological force. We live on land and sea, in the air, and even in space. How did we transition from small, probably nomadic, populations in some corner of Africa to the dominant global species? What story can we tell our children about the human family's origin, its diversity, and its compulsive tendency to migrate? Perhaps we should begin with the importance of climate.

CLIMATE CHANGE HAS BEEN THE MAJOR DRIVER OF HUMAN MIGRATION

Climate and climate change are key determinants of life's possibility and trajectory. Therefore, it is not surprising that changes in climate have been and continue to be a major factor determining when populations migrate. Our nomadic hunter-gatherer forebears were not adventurers or explorers. They were shoppers, seeking food, and sometimes they were refugees leaving good places that had turned bad. Climate determines when and where nature's market stalls will be open and stocked and what will be on offer. Nomads learn that berries, ripened by long, warm days, can be found here and that antelopes feeding on grass that has grown during the rainy season can be hunted there. They know when swollen spring rivers will have runs of easily caught fish. Changes in climate can disrupt the weather-driven rhythms of all human societies but especially and immediately those of hunter-gatherers who have no fixed dwellings to attract them to linger. The San, a nomadic people of Southern Africa, build quick temporary shelters of branches and leaves that do not long outlast the departure of their builders, so that if a band returns to the same place, it is like a new beginning, to be abandoned again when the local resources are exhausted.

Our origin just happens to fall within the only ice age of the last 250 million years. Despite the name, ice ages are not uniformly frigid but are marked by periods of cooling when ice sheets advance and times of warming, called interglacials, when the ice retreats. We are living within one of the periodic warm spells of the Quaternary Ice Age, the name given to the ice age that began about 2.6 million years ago. If we are residents of Maine, Moscow, or Southern Madagascar, from time to time the ground beneath our feet was under a glacial ice sheet a thousand or more feet thick. Since ice sheets lock up water, more than landscape and temperature are affected.

Periods of glacial advance are cooler and drier. Sea levels fall, lakes diminish or disappear, and drought becomes common, causing forests and grasslands to shrink, some of them becoming deserts. Falling sea levels can connect land areas separated from each other by bodies of water, allowing animal populations confined to one land mass to migrate onto and populate others normally inaccessible.

In contrast, during interglacial periods, temperatures rise, ice sheets melt, water is released, sea levels rise, coastal areas become sea beds, and more rain falls. Increased rainfall transforms arid areas and even deserts into verdant landscapes, lush with plants and populated by the animals that move in to exploit them. For the 300,000 years or so that modern humans have existed, climate change has been a factor, sometimes providing opportunities but often presenting challenges, some existential, that had to be met, overcome, or accommodated. During this time, climatic conditions underwent periodic moderations that produced vegetated corridors rich in plant and animal foodstuffs between Africa and the Arabian and Sinai Peninsulas. These corridors invited migration from Africa onto the Eurasian landmass.

A vivid, instructive, and relatively recent example of the impact of climate change on the landscape, population, and culture of a region is provided by Northern Africa. Twelve thousand years ago, the pattern of rainfall made the Sahara a verdant region of grasslands and forests with many lakes and rivers. As expected for an area rich in primary food resources, it supported thriving populations of *H. sapiens*. Then around 5,000 years ago, patterns of monsoon rainfall shifted, greatly reducing rainfall, and vast areas of North Africa became arid, so much so that the area we call the Sahara again reverted to desert. One can imagine that once the consequences of lower rainfall were perceived as enduring characteristics of the area, those humans who could migrated to greener fields, and the stream of newcomers to the area dried up.

THERE HAVE BEEN MANY WAYS OF BEING HUMAN

So far, we have ignored the fact that we are not nature's only version of human. We are not even her first. Humans are members of the genus *Homo*, and within the genus *Homo* there are many species; all except ours are now extinct. One of these is *Homo erectus* ("upright man"), arising in Africa about two million years ago. These early humans were the first of our genus to migrate and spread as far east as the part of Asia we now call China. Another is *Homo neanderthalensis*. The Neanderthals were similar enough to us that a casual observer might have trouble distinguishing members of this large-boned and more heavily muscled species from a Philadelphia Eagle or Dallas Cowboy. In contrast, the facial features and skull conformation of a representative *H. erectus* would be easily distinguished from ours. So much so that if he were in an airlines' middle seat, both his window and aisle seatmates, though pretending not to, would glance his way, repeatedly, and perhaps a bit anxiously.

The cultural artifacts associated with fossilized bones of *H. erectus* let us know that they made a variety of tools and used fire in controlled ways. Maybe they even cooked. Whether or not some of these earliest humans were chefs, they hold the record for longevity. Most students of human evolution think *H. erectus* arose in Africa two million years ago and became extinct only about 100,000 years ago. That's a very long run. Somehow, this species of early humans managed to survive the weather extremes of not just three or four glacial advances and retreats, they endured dozens of them. Our brief eulogy for this probable progenitor of us and the Neanderthals has to include an admiring recognition of the adaptability, resourcefulness, and sheer toughness (*resiliency* is too weak a word) of a species that could occupy so many different environments, maybe not quite up there with the cockroach, the rat, and us, but still impressive.

HUMANS LIKE US HAVE BEEN AROUND FOR MORE THAN 300,000 YEARS

While it is not possible to pinpoint a time of our origin, it was pushed back in 2017 from earlier estimates of around 200,000 years to more than 300,000 years ago. Discovery of fossil remains of early *H. sapiens* in a cave at Jebel Irhoud, a Middle Stone Age archaeological site in the North African country of Morocco, revealed that our species is much older than previously thought. The artifacts found with these fossils and their immediate environment have shown that they were crafting tools and using fire. Up until this discovery, the earliest fossils of modern humans, all more than 160,000 years old, were found in East Africa: two skulls in the Omo River Basin of Ethiopia, and three skulls—two adults and one child—found at Herto on the Awash River in northeastern Ethiopia. Recent discoveries have also revealed modern human fossils in caves of the Mount Carmel region of Israel that are from 100,000 to at least 170,000 years old, demonstrating that there were human migrations out of Africa long before the one that took place more than 50,000 years ago and that populated the rest of the world. The findings in North Africa have forced the time of human origins in Africa back to at least 300,000 years ago. If a fossil of that age is found, it is unlikely to represent the very earliest creature of its kind and it is probable that our origin was significantly earlier than this 300,000-year benchmark.

While each of these fossil finds is just an isolated snapshot of parts left by one or a few individuals of a much larger population, it is still possible to learn a great deal from their study. The size of the cranial vaults allows inferences of brain size and the shape of the skull, facial contours allow modeling of facial features, and studies of dentition in these skulls allow us to learn something about their diet. These skulls came from people who had brain sizes that matched or exceeded contemporary averages, and facial reconstructions would indicate they looked much like someone in the crowd. Dentists in the local family practice would find the teeth of these ancestors to be the same size and arrangement as those of their clients but almost certainly in need of more diligent flossing.

Migration within Africa was a prelude to migration out of Africa. *H. sapiens* spread from its probable origins in North Africa to eventually occupy the entire continent. This would have taken a while. Africa is a very big place, around 5,000 miles north to south and 4,600 miles east to west, an area large enough to contain India, China, Spain, France, Germany, Greenland, and the United States and still have space left over to accommodate little Switzerland and tiny Luxemburg. Nevertheless, by 120,000 years ago, modern humans had spread throughout Africa, populating regions all the way down to the Cape of Good Hope, and migrating westerly and northwesterly across and up through the region we now call the Sahel, across the Sahara, and all the way to the northern rim that lies along the Mediterranean.

These were major migrations and, along the way, they forced adaptation to a great variety of environments: temperate and equatorial; rain forest and savannah; desert and flood plain. This diversity of environments required an answering diversification in the culture of Africa's people as they spread around and over the continent where some remained, and later, others would leave, emerging from this large and geographically varied landmass to explore, exploit, occupy, and transform the larger world beyond. In Africa and outside it, they underwent a dazzling degree of cumulative genetic changes in response to the selective pressures of different environments and the relentless and kaleidoscopic exchange and shuffle of genes within and among so many human populations. This process of gene shuffling and diversification has been taking place in Africa longer than anywhere else on Earth. This has made the populations of Africa the most diverse on Earth, and in Africa, as elsewhere, the perception of difference has prompted multiple forms of local racism, usually appearing under the alias, tribalism.

HUMANS HAVE BEEN SMART, BUSY TOOLMAKERS FOR A LONG TIME

Almost from the start, Africa's rising population of modern humans appears to have had many of the traits and behaviors we would associate with today's humans. A vivid example of this comes from the Pinnacle Point archeological find. This was made in South Africa during construction of a golf course near the town of Mossel Bay. When caves were discovered in this seaside town in South Africa on the Indian Ocean not far from Cape Town, exploration of these caves revealed a record of habitation beginning as early as 165,000 years ago. The investigators concluded that the area's ecology would have afforded hunter-gatherer societies a great diversity of plants and ample sources of animal protein. The vegetation included plants distinctive for such energy-rich underground storage organs such as tubers and bulbs (think of potatoes and onions). These could have been important contributors to the diet. Notably, some of the tubers and root vegetables that dominate in that region of South Africa are well-suited to the diet of hunter-gatherers because their lower fiber content makes them easily digestible, a feature of particular significance for child nutrition. Regarding animal protein, although this coastal location was not a good spot for hunting large land animals, marine mammals were there for those with the technology and skill to harvest them. We know that these early foragers developed the requisite skills because there was clear evidence that their diet included seal and even whale meat. Importantly, the caves provide evidence that the people of Pinnacle Point also exploited the area's abundant populations of shellfish.

However, it turns out that harvesting the mussels and sea snails of these waters from their rocky intertidal habitats was not easy. For much of the year, the surf is quite dangerous, and collecting would have been perilous. Only during the spring when calm, low tides predominate would collection have been relatively safe. Archeologist Curtis Marean, a pioneering investigator of this site, has speculated that these early shore people probably worked out something like a simple lunar calendar that would have allowed them to anticipate the arrival of spring and its permissive tides, enabling scheduling of safer coastal harvests. On the other hand, maybe they just got up in the morning, looked at the surf, saw calm conditions, and waded in to collect dinner.

Other technologies, one quite advanced, were associated with populations that inhabited Pinnacle Point. In addition to the manufacture and use of a variety of conventional stone tools, the inhabitants also made small, thin, sharp bladelets. These could be attached to a shaft and, when they were damaged, replaced, thereby conserving the shaft of the projectile. A surprising and intricate process was used to manufacture the bladelets at the heart of this early version of "plug-and-play" technology. The bladelets were made from a type of rock we know as silcrete, which, in its native form, is incapable of yielding sharp, strong blades—unless it is heated. By 72,000 years ago, the people at Pinnacle Point had worked out an elaborate process for the production of sharp silcrete bladelets. Studies have pieced together an early process of engineering that involved a number of steps that had to be executed in a particular way and in a precise order. Sandpits for controlled heating of the silcrete were built, and fire was used to gradually bring the temperature of the silcrete to well above the boiling point of water, hold it at that temperature, and then slowly reduce the temperature (a process we would call annealing) to produce the blades. Because this process was passed down through generations of inhabitants, these people had to develop what might be thought of as a sort of vocational training. The ability to develop and transmit complex, multistep technologies is a clear demonstration that the advanced cognitive and social behavior we associate with *H. sapiens* was present early. Indeed, study of the hunter-gatherer populations who inhabited the caves at Pinnacle Point demonstrated that the assembly and use of many elements of the distinctive modern human "cultural package" have been around for a long time.

Archeological discoveries such as those at Pinnacle Point confirm that impressive technologies appeared early in our long African residence. Finding, dating, and studying cultural remnants at sites occupied by humans—the tools, animal bones, shells, and trash—let us know something about their activities of daily living and their skill levels. Examination of some of the tools, implements, and weapons recovered from various sites reveal a high level and wide variety of skills. Archeologists have also been impressed by the finding of beads, shells, and markings that have no apparent utilitarian function but may have been selected, created, or used for purposes that could be regarded as symbolic or aesthetic. For example, ochre, a mineral conferring shades of red or yellow on stones containing it, was in use as a coloring agent over 160,000 years ago. Beads and shells colored with red ochre have been found at more than one site. In some instances, features of the shells strongly point to their collection for purposes other than food. In some cases, it is apparent that the amount of meat that could be obtained from the small shells employed was too insignificant to justify the effort and possible risks of harvesting them just for food. Consistent with their use as decorative items, perhaps as components of jewelry, there are instances in which these shells have been punctured, as though to string them together, perhaps for use in a necklace or bracelet.

Artifacts made of stone and shell have the virtue of being able to last for very long periods of time, for tens of thousands of years. We also realize that if these distant ancestors were anything like us, in addition to stone or shell, they would have made a variety of other objects from materials that are less permanent. These would include wood, feathers, easily eroded or broken pottery, and vegetable dyes that would wash away or fade with prolonged exposure to sunlight and the elements. Of course, we have no way of looking back tens of thousands of years and examining creations that don't resist time. However, we feel certain such perishable and pretty things were made, enjoyed, and perhaps, along with their artisan creators, cherished and admired.

THERE WAS A TIME WHEN MODERN HUMANS
TEETERED ON THE BRINK OF EXTENSION

The Skin of Our Teeth, the title of Thornton Wilder's classic play, is an apt descriptor of a point during our long and exclusive residence in Africa when the human population suffered a catastrophic decline and was reduced to a small population of only a few thousand, possibly just hundreds of individuals, perhaps living in the same general area. This was a perilous time, and failure to pass through this population bottleneck would have resulted in the extinction of modern humans. Clearly, there was a different and far less grave outcome. However, all seven billion people on Earth today arose from the small population of humans who made it through this existential population bottleneck. We are the children of a relatively small number of Adams and Eves.

Our first indications that this is the case came from studies of the DNA found in mitochondria, subcellular units that play essential roles in the conversion of energy found in foods to chemical forms that power many essential life processes. Early studies produced the unexpected and striking finding that diverse human populations, different nationalities, and groups who would certainly be labeled by some as different races, bear similarities in their mitochondrial DNA that can only be explained by an origin from a very small population of common ancestors. Moreover, since one inherits all of one's mitochondria and mitochondrial DNA from the mother and not the father, we all, at least figuratively, can trace an ancestry back to what has been dubbed "mitochondrial Eve." Actually, we can trace ancestry back to a small population of women whose genes passed through the nearly catastrophic population bottleneck just described.

WHAT WAS DAILY LIFE LIKE IN
HUNTER-GATHERER POPULATIONS?

What was daily life like for our ancestors during those early times? Until about 10,000 years ago, human subsistence followed the pattern of many other mammals, with variations depending on the local environment, a pattern called "hunting and gathering." Although human consumption has tended to include both vegetable and animal foods, many other species are more specialized in their diets. What this means is that our ancestors killed and ate wild animals and gathered and ate wild edible plants, without having the technology to breed or cultivate them. They probably followed the pattern seen in many contemporary hunter-gatherer populations where hunting and fishing are done by males and the gathering is done by females. Horticulture and animal husbandry did not exist, and the key technologies were those of hunting, transporting, and handling food—the focus was on the fabrication of weapons, tools for food processing, and containers for transporting food and gear. In some places, particularly where animals grazed, drinking water was nearby and plant growth was diverse enough and the climate stable enough for edible plants to be gathered year-round, a relatively settled existence was possible.

However, in many places, a hunting-gathering adaptation meant living in a series of temporary sites until the low-hanging fruit had been gathered or available supplies of water or firewood were exhausted. In this context, migration simply meant moving on without a defined destination—walking into the unknown with few possessions, since everything, including young children, had to be carried. Bands were small. A couple with one or two children might even travel alone for a period, or an all-male group of hunters might follow a wounded animal, but humans learned early on to keep track of kinship.

Virtually every human society we know of pays some attention to kinship as it connects living members of the community, partly to reinforce mutual help, and partly to avoid marriages regarded as too close in that particular society. Almost every society has a theory, passed on through the generations, of its own origins. It is not surprising, then, that most individuals have a set of ideas about their own ancestry. Contemporary studies of genetics sometimes ratify those ideas but increasingly supply additional—sometimes surprising—information. Indeed, as we have seen, a flood of new discoveries and insights since the start of the millennium has forced a revision of what most of us thought we knew about the history of our species.

Returning to our contemporary San hunter-gatherers, there were traditional times and places when larger numbers would gather to exploit a particular resource, and marriages would take place, usually by a man from one band marrying a woman in another band and joining that band as an additional hunter—doing what anthropologists call bride service. Other things were exchanged as well—connections that promised help when needed were more important than property, as were tools and ornaments, such as ostrich eggshells to carry water or bags fashioned from skins or woven from whatever reeds or sticks might be available and passed easily from hand to hand. Occasionally, bands would clash, and in some instances, strangers might be regarded as game, but the organized warfare we associate with settled societies did not exist.

The once superior virtue of a nomadic hunter lifestyle as compared to a settled lifestyle is captured in a myth "that never happened but is always true": the story of the brothers Cain and Abel. As the story goes, two brothers adopted different ways of life, which led to them making different kinds of offerings to God: one offered the produce of his fields, the other offered the meat of animals killed in the hunt. As the story goes, God disdained the vegetable offerings of Cain, the farmer, and accepted the flesh offerings of Abel, the hunter; Cain then murdered Abel. Many people, vaguely remembering the story, assume that Abel was the farmer and Cain the hunter (the "meat-eater"), but it was Abel who was the hunter and whose offering was preferred. The myth reflects a time before the balance of power had tipped to settled living from the migratory patterns of hunter-gatherers. It presages the generally greater power of settled farming societies compared to migratory ones sustained by hunting and foraging.

In today's world, though dominated by settled societies, there are still groups of hunter-gatherers in South America, in the Pacific Islands, and in Africa. However, as neighboring populations have established populous settled societies with complex and powerful technologies, members of hunter-gatherer societies have been regarded as inferior and have been exploited, even enslaved, and excluded from intermarriage. One of the distinctive characteristics of any society is its preference for marrying in (called endogamy) or out (called exogamy). Since endogamy tends to preserve and perpetuate distinctive characteristics of a breeding population, most mating systems include some degree of endogamy.

NISA'S STORY

Although fossils tell much, the rhythm and tone of day-to-day social interactions are not part of the fossil records. We have to guess. However, maybe our guesses may be informed by an examination of the lifestyles of contemporary hunter-gatherer cultures. A revealing look at the cycle of life experienced by a San woman is presented by Marjorie Shostak in her book, *Nisa*, an intimate look at the day-to-day life of a member of a migratory culture that lived mainly by hunting and gathering, a tribe of the San.

If Nisa's mother followed the tribal practice, when she sensed that the time of delivery was near, she probably would have told some women in the group that she was going to such-and-such a spot, perhaps a mile or so from the campsite. There, she experienced labor and gave birth alone, returning with the newborn Nisa. The child's first food was breast milk and then, as time passed, food pre-chewed by her mother and fed to her by mouth-to-mouth contact in an act somewhat akin to "French kissing." Because very young children cannot chew and commercial baby foods were, of course, unavailable, this was an effective way of adding a wider range of foods to Nisa's diet. Throughout history, premastication has been widely practiced and is still employed in some parts of the world today. Getting back to Nisa, breast milk would continue to augment her diet until she was perhaps three years old, perhaps even four. Therefore, it is reasonable to surmise that infants born to our earliest *H. sapiens* ancestors would have been fed in ways that parallel those of the !Kung. It has been widely recognized that in the industrialized world women now mature earlier sexually, and that was the case in the earlier history of *H. sapiens*, probably because of better diets and reduced activity.

Thus, in many industrialized societies, girls now experience their first menstruation at around eleven or twelve years of age and are capable of conception soon thereafter. Among the San, puberty and first menstruation occur on average at fifteen years of age followed shortly by marriage, which begins with a few years during which conception is apparently difficult; first childbirth is not likely to occur until eighteen to twenty-two years. As mentioned above, infants are not fully weaned for up to four years, and continued frequent nursing appears to inhibit conception as long as it lasts. Menopause occurs sometime between the ages of forty and fifty, and women average 4.7 pregnancies in a lifetime, with a survival rate to adulthood of 50 to 60 percent. Adult life expectancy as hunter-gatherers is about sixty-five years. This works out to a roughly stable population, with an average life expectancy into the sixties with fairly good health. Nisa herself lived into her sixties.

In this hunter-gatherer population, a woman approaching menopause may adopt the child of a relative. Adults commonly enjoy and pay attention to all children—it is literally true, in the often quoted African proverb, that "it takes a village to raise a child," but where settlements are temporary, residence is evanescent, and shelters are built for only short-term use from available materials, it makes more sense to say that "it takes a community to raise a child." Everyone is involved and aware. Elders are respected, valued, and assigned special roles—they shape the arrows with which the young men hunt, distribute the meat, and participate as healers in community healing ceremonies. Virtually everywhere, settled living with domesticated animals involves a new set of health hazards, a result of denser populations, accumulated dirt and bacteria, more concentrations of insects and rodents, and the infecting parasites they bring. In its original form, the San lifestyle was a healthy one, although health has deteriorated with increasing contact with outsiders and animal herders. The nomadic lifestyle of the San aligns with the larger migratory history of humans.

Even before the origin of modern humans, the group that includes all humans now living on Earth, earlier branches of our family were on the move. As mentioned earlier, the earliest human, *H. erectus*, appears in the African fossil record around two million years ago, spreads through Africa, doesn't remain in the ancestral home but migrates into Europe, throughout the Near East, and into Asia, establishing a clear presence in China by 1.7 million years ago. *H. erectus* was the first human out of Africa, and it is quite possible, even likely, that over the long span of hundreds of thousands of years, not just one but successive bands of *H. erectus* left Africa. It is also possible that some of these migrants, or their descendants, might have come back. However, it is certain that at least one or a few populations sustained a prolonged existence outside Africa, pushing into new territories and becoming established there. Almost certainly, even before the *H. erectus* line died out tens of thousands of years ago, many earlier lines died out. However, some evolved. *H. erectus* or more advanced transitional forms gave rise to Neanderthals, large-brained humans that had established populations in Europe by 430,000 years ago.

DNA SEQUENCING PROVIDES REVOLUTIONARY INSIGHTS INTO THE NATURE AND VARIETY OF OUR ANCESTORS

From DNA sequencing of entire human genomes, we know that Neanderthals and *H. erectus* were not the only two early human populations living outside Africa before the origin of *H. sapiens* (modern humans). The first sequence of a complete human genome was announced in 2001. Since that landmark event, the technology has advanced rapidly, and it is now possible to determine a genome sequence with very small amounts of DNA. Surprisingly, it is possible to extract DNA from human fossils tens of thousands of years old, and this has opened new approaches to the study of human evolution.

In his revelatory book, *Who We Are and How We Got Here*, David Reich, a pioneer in the application of DNA sequencing to the study of ancient human genomes, shows how the collaboration of DNA sequencing with physical anthropology and archaeology has dramatically revised and added to our understanding of human prehistory. Fossils provide tangible physical evidence of a creature's existence, paleogeology can provide estimates of the time the fossil was deposited, and archaeological studies can say something about artifacts found with the fossils, relating them and the context in which they were found to other artifacts and contexts. Anthropology helps determine, interpret, or construct an interpretive fabric of culture that helps understand behavior. However, fossils are usually incomplete, such as sometimes only a few teeth embedded in a fragment of a jawbone. Nevertheless, a great deal can often be divined about what the donor of the jawbone fragment ate, how old he or she was, and the likely size and shape of a jaw, and from that, broader inferences about body structure can be made. However, fossils don't directly tell much about the ancestors of the donor, and if they are bone, they don't directly inform us about the detailed physiology of the donor.

In contrast, a DNA sequence obtained from just a single tooth or appropriate bone fragment can provide information about many physiological features, including aspects of the nervous system. It can also provide information on many anatomical features, including eye color and skin color. Furthermore, deep insights into the detailed ancestry can be deduced reaching back into the very distant past to conclude that Neanderthals were definitely in the family tree of modern humans, even offering some estimate of how long ago and to what extent these archaic humans made contributions to lineages of modern humans. It is not an overstatement to say that the addition of DNA sequencing to the study of human origins has revealed much that was not and could not be known from the application of hitherto conventional approaches. However, it would be a serious mistake to assume that genomic sequencing has replaced archeology, paleogeology, and physical anthropology in the study of human evolution. It has been an extraordinarily powerful complement to these traditional methods, enabling studies at a depth and precision impossible before its introduction. However, it requires these older methods for the indispensable provision of fossils and context. With this understood, consider one of genomic sequencings impressive recent discoveries, which is outlined in the following account.

East of the Neanderthals' European home ground, a fossilized bone that was once part of the little finger of a child was found in southern Siberia's Altai Mountains in the Denisova Cave. This fossilized little finger led to a big finding. Although the bone fragment had waited tens of thousands of years to be discovered, it was still possible to recover some DNA and sequence it. Studies of the sequence revealed that it matched no human species previously examined. It was very likely a new species of ancient and now extinct human, different from modern humans, different from *H. erectus* and different from, but most similar to, Neanderthals. The new variety of human was named Denisovan and appears to have last shared a common ancestor with Neanderthals more than 500,000 years ago. As with the Neanderthals, Denisovans may have ancestral ties to Africa through *H. erectus*. All of this makes it clear that before our arrival in Europe and Asia, the Eurasian landmass already had a long and varied history of human activity.

MODERN HUMANS MIGRATED OUT
OF AFRICA MORE THAN ONCE

Periodic cycles of warming and greening generated corridors from time to time that provided humans occupying the northeasterly parts of Africa the opportunity to launch migrations out of Africa and onto the Eurasian landmass. Eventually, seizures of these opportunities led to at least two, and possibly more, waves of humans migrating out of Africa. An early migration more than 100,000 years ago was not very successful and failed to establish an enduring population in the rest of the world. It was a later one 50,000 to 60,000 years ago that eventually populated and established permanent settlements of modern humans in all parts of the Earth.

However, we must be aware that the great migrations that peopled the planet were not the playing out of some grand scheme hatched by a council of distant ancestors. In most cases, the migrants would not have viewed themselves as immigrants, consciously choosing to relocate to new lands. They were simply hunter-gatherers trying to scavenge enough from the land to feed themselves. They went where they thought food might be and returned to places where they had found it. Conditions of climate that favored abundant sources of edible vegetation and the game that made up the food chain anchored by this plant life made the dwell time of our nomadic ancestors longer in food-rich areas. On the other hand, they spent little time in areas, such as deserts, where plants, animals, and water were sparse.

Social interactions within and between groups must have played roles in determining who moved on and when. No doubt some bands sought to monopolize particularly desirable areas for their own exploitation by aggression against competitors. Also, perhaps from time to time, because of internal dispute or amicable agreement, a band would decide to split up, each faction going its own way. One imagines that resource availability was a frequent and often decisive factor in these disputes and decisions. Because climate is a key determinant of resources, it surely played a role in creating the stakes that drove these negotiations. Whatever the constellation of causes, some bands of humans took paths north and eventually into the part of west Eurasia we call Europe. The way this trek played out provides a dramatic and quite surprising illustration of the degree and rapidity of the change migration and genetic mixing can have in shaping the populations that ultimately occupy a region.

THE "WHITE" PEOPLE OF NORTHWEST EUROPE
ARE ONE OF NATURE'S RECENT INVENTIONS

Today, when DNA sequencers compare the DNA of European populations from north to south or from east to west, they find some small degree of difference. However, the variation between the DNA of a typical European population and a typical Han Chinese population is six times greater than the difference among Europeans. Similarly, members of the Han Chinese population differ from each other much less than they differ from Europeans. Today, a traveler touring Northwest Europe and stopping in England, Germany, and France sees white people—populations that are generally light-skinned, with black, brown, red, or blond hair and most with blue, gray, or light brown eyes. However, in *Who We Are and How We Got Here*, David Reich tells us that prior to 6,000 or 7,000 years ago, a traveler visiting those locales would have seen dark-skinned people, some of whom were blue-eyed. Regarding occupation, none of them would have been farmers. They would have been hunter-gatherers.

A visitor to England 6,000 years ago would have had the opportunity to visit the site where Stonehenge, the largest stone monument in the world at the time, was being built. These builders and the onlookers at the construction site would have been a dark-skinned population, typical of Northwest Europe at the time. Just 1,500 years later, long after their completion of England's iconic Stonehenge, that population would have disappeared, replaced in England and in other parts of Northwest Europe by the light skins, light eye shades, and mixture of hair shades recognized as iconically "White," a look most of us have assumed had been typical of the area for a very long time.

This surprising conclusion comes from a survey of DNA sequences obtained from individuals living in a window of time spanning a period 3,000 to 10,000 years ago and collected from humans that lived in a broad swath of Western Eurasia. The specimens examined came from an area that runs from the Arctic in the north down to Turkey, the Fertile Crescent, and Iran in the south and stretches from the Atlantic in the west to include portions of the western Asian steppes in the east. With archaeology and anthropology playing indispensable supporting roles, it is the determination of the sequence of ancient DNA that leads to the striking realization that "White" Northwest Europeans are one of nature's surprisingly recent inventions. What follows is how Reich, a major contributor to our knowledge of ancient human genomes, and his colleagues have outlined the story of how West Eurasia became Europe.

We start with seed populations from three areas. Northwest Europe was the redoubt of dark-skinned hunter-gathers that had not yet developed agriculture. To the south and east, in areas we would now designate as parts of Turkey, the Fertile Crescent, and Iran, light-skinned populations of farmers were practicing the agriculture and animal domestication that had begun in that part of Eurasia more than 9,000 years ago. Their cultivation of a sustaining "package" of grains including wheat and barley, along with the husbandry of sheep, cattle, and hogs, gave them a base for expanding into much larger populations than could be sustained by hunter-gatherer lifestyles. Eventually, these farming societies and their agriculture expanded into the north and west of Eurasia, mixing with and absorbing the hunter-gatherer cultures already there. To the east, on the West Asian Steppes, there was the Yamnaya, a pastoralist population that maintained herds of cattle and sheep and had learned to domesticate and use the horse.

At the edge of the Steppes, possibly in the region of the Caucasus, the wheel had been developed. This technology was also adopted by the Yamnaya, who eventually made a fateful coupling of wheel with horse. This was transformative. Hitching wheeled carts to horses gave them mobility, speed, and a capacity to carry that encouraged migration and enabled movement for purposes benign or hostile at speeds greater than previously possible. This package was likely the key factor behind their successful sweep into the part of Western Eurasia we call Europe and, as mentioned later, south and east into what is India and Pakistan today. Yamnaya artifacts, such as huge ax-like weapons, make one doubt that confrontations with the Yamnaya were always settled peacefully. DNA sequencing provides definitive proof of the enduring contribution of this Steppe population to the complex gene pool from which the much more homogeneous one we now see in Europe emerged.

EASTERN MIGRATION OF HUMANS ACROSS THE NEAR EAST TO SOUTHEAST ASIA AND AUSTRALIA AND NEW GUINEA, TOO

Leaving what would become Europe, our narrative returns to that greening land corridor and the still-narrowed Red Sea that encouraged and facilitated a move from Africa to consider another direction migrants may have taken in search of edible plants and in pursuit of a variety of animals. Because the fish, shellfish, and crustaceans found at the seashore would have provided an abundant source of easily harvested animal protein, easterly migration along the shorelines would have been especially attractive. Migrants who brought with them or developed a cultural toolset that included successful technologies for fishing and for trapping and harvesting other aquatic species would have been tempted to move along or close to shorelines and their bounty. Although other populations also found their way into South Asia, generations of shoreline migrants, shedding salients that would stay or take other directions, likely found their way to South Asia, where 50,000 years ago there was still so much water tied up in the world's glacial ice sheets that the finger of Southeast Asia on which today's nations of Thailand, Cambodia, Vietnam, and Malaysia sit were joined to a landmass now called Sunda, which incorporated the area that is now the island nation of Indonesia. Just across from Sunda over a relatively narrow body of water lay another landmass called Sahul that incorporated both New Guinea and Australia.

At that time, there was a network of islands so close together that one might very well have been able to cross the bay from Sunda to Sahul, moving from island to island, without ever losing sight of land. This would have enabled migration from South Asia to what would become Australia and New Guinea without having to deploy seafaring technologies that simply did not exist at the time and would only appear much later in human history. Subsequently, water liberated by the recession of the glaciers would reestablish higher sea levels, again separating Indonesia from the South Asian landmass and dividing Australia from New Guinea. The modern humans who occupied Australia and New Guinea 45,000 to 50,000 years ago would live hunter-gatherer lifestyles throughout most of their long history on these vast islands. Because they lacked well-developed agriculture, the calories required to build and sustain large populations were also lacking. We see very different outcomes on these isolated landmasses from those seen where the introduction and spread of agriculture fueled population explosions and the development of settled living.

A CLOSER LOOK AT THE INDIAN SUBCONTINENT TELLS A SURPRISING STORY

Today, the Indian subcontinent and the nearby island, once known as Ceylon, hosts the nations of Pakistan, India, and the island nation of Sri Lanka. A look at the Indian subcontinent reveals the formation by migration and mixture of many populations. Like the European story, the Indian story is a complicated one with the migration of populations from outside the subcontinent arriving and, to varying degrees, intermixing with the populations they encountered there. The short version of a much longer, much richer, and still unfolding story is that the peoples we see in the subcontinent of today arose from a diverse and varying mixture of several different populations including hunter-gatherer populations who had been living on the land for tens of thousands of years before the appearance of settled agriculture along the Indus River. DNA sequencing has identified two major groups from which genetically mixed populations of today's Indian subcontinent arose.

There was an ancestral darker-skinned southern Indian (ASI) population and an ancestral lighter-skinned northern Indian (ANI) population. Despite the names, a precise partition of skin color does not geographically divide India. Indians, many of them darker skinned whose DNA sequences would draw more from ASI ancestry, are on the streets of northern Indian cities such as New Delhi, and lighter-skinned Indians, often having more ANI ancestry, attract no special notice in southern Indian cities such as Bangalore or Madras. If we look into the origins of the ANI and the ASI, a partial story of their history emerges. Comparative studies of DNA sequences show that populations from Iran, probably agriculturalists, contributed to the gene pools of both the ANI and ASI. Culturally, these populations contributed to the cultivation of wheat and other crops. Perhaps a quarter of ASI ancestry came from Iranian farmers and most of the rest from hunter-gatherer populations. About half of ANI ancestry was Iranian, and most of the other half came from the Yamnaya, the highly mobile horse-cart pastoralists we met in the discussion of Europe.

However, it would be a mistake to view the people of the subcontinent as simply lineal descendants of these two ancient founder populations. In the east, there is readily detectable ancestry from East Asia. There is much more to the story than two largely homogeneous populations in India, freely mating within their own populations. Rather than homogeneity above and below certain latitude, we see a gradient of mixing of these populations as we move from north to south. Perhaps one of the most important bars to free intermixture within a large fraction of the Indian population is the caste system practiced by a large and influential segment of the Indian population.

The caste system was introduced long ago and, though now outlawed, is still highly influential in dividing a significant fraction of Indians into many smaller populations that marry within their own group but are culturally restrained from intermarriage with other groups. Furthermore, the division of large numbers of Indians into the Moslem or Hindu faiths draws a major fault line between peoples of the subcontinent. For all of these reasons and so many more, India remains a place of great genetic complexity and diversity.

STUDIES OF ANCIENT DNA FROM EAST ASIA REVEAL THE ORIGINS OF THE CHINESE PEOPLE AND A TIGHT CONNECTION BETWEEN KOREA AND JAPAN

East Asia has been occupied by humans for more than 1.7 million years. There was a human presence there when the "out of Africa" wave of modern humans arrived 45,000 to 50,000 years ago. Around 9,000 years ago agriculture had begun independently at two different sites. At one of these, along the Yellow River, millet was domesticated, and to the south, along the Yangtze River, people learned how to grow rice, a grain that would become a key and transformative calorie-rich addition to the diet of cultures that developed and adopted it. The domestication of rice was epochal, and even today, rice is the source of 20 percent of the calories consumed by humans worldwide. The Han, the ethnic group that united China and who now number over 1.2 billion, comprise the largest ethnic group in the world and emerged from these early agricultural centers. Among some of the present-day Han, there is a perception of some difference in the northern and southern Han population.

The reality and depth of this difference will be definitively determined by increasingly comprehensive comparative genome studies now being conducted by Chinese researchers engaged in a massive effort to build a comprehensive picture of the current and historic genetic structure of the Chinese population. If a north-south genetic gradient in the Han is demonstrated, perhaps the intersection and intermixing of the northern Yellow River agricultural populations and Yangtze River agriculturalists to the south might be a significant contributor to the formation of such a gradient. However, as is the case with so many populations, studies of the Han genome will surely point to a rich history of genetic interchange with other populations who have been absorbed by history and whose existence will be known only by the genetic signatures they left in successor populations. It is already known that the Han, like other East Asians and Europeans, also bear genetic traces of Neanderthal ancestors.

Beyond what is now China, we also see the spread of agriculture and agriculturalists into Southeast Asia and into Korea and subsequently into Japan. Farming populations from the Yangtze region spread into the parts of Southeast Asia where we find Vietnam, and there they mixed with resident hunter-gatherer populations. Farther north, the Korean peninsula absorbed the agricultural advances developed in China and a rich farming culture developed. This was a precursor and probably an enabler of the creation of a highly sophisticated technology of pottery making that was one of the great achievements of early Korean culture.

Eventually, Korean agriculturalists crossed the water separating the peninsula from Japan. There they found hunter-gatherer populations, some of which were the ancestors of present-day Japan's small Ainu population, who had lived a foraging existence for tens of thousands of years. With the arrival of Korean farmers, things changed, and the hunter-gatherers were displaced and largely absorbed. Although we do not know the details of this history, today we see a Japanese population whose DNA sequences reveal about 80 percent Korean and 20 percent hunter-gatherer ancestry. Since this early episode of extensive genomic intermixing, Japan has not hosted substantial in-migration of other populations for millennia and is famously perceived as a homogeneous population. However, as DNA-based genetic studies clearly show, they are not a "pure" line of humanity. Japanese, too, are a mixture of ancestries, genetic "mutts," as most other populations have proven to be when looked at through the discerning lens of genome sequencing.

MODERN HUMANS LEAVE NORTHWEST ASIA AND MIGRATE TO THE AMERICAS

By 25,000 years ago, humans were established in areas of Northwest Asia we would today recognize as Siberia. However, today's warmer climate would be different from conditions at that time, and the watery expanse of the Bering Sea would be unfamiliar. Although the last glacial advance was turning to a retreat, there was still enough water sequestered in ice sheets for sea levels to be much lower. Where the Bering Sea now lies, there was a land expanse, dubbed Beringia, connecting Asia to North America. In some places, this irregular landmass was almost 3,000 miles west to east and 1,000 miles from north to south. Some Asians left Siberia 25,000 years ago and began a habitation of Beringia, then living and walking on what is now underwater. Over a period of perhaps as long as 10,000 years, these original settlers and later immigrants hunted and gathered on this now largely submerged land before moving onto Alaska, quite unaware that they had become the first Americans.

From that northerly starting point they raced down the Pacific Coast, reaching South America, and continuing down its coastal regions. Within a relatively short time, perhaps as little as a thousand years or so, they had reached and left cultural artifacts in areas we now call Peru and Chile. The current hypothesis is that this rapid spread from an Alaskan beachhead was not all made on foot but that some form of watercrafts were used, perhaps avoiding geographical obstacles and speeding the pace of the southern migration. At one or more times during the southerly progress of these people down the coast, salients took easterly turns and spread across the continent, reaching the Atlantic along the way, and after arrival, they became the Apaches, the Sioux, the Iroquois, the Cherokee, and uncounted others, known to themselves but not to us.

Here in the Americas, people did what they have done in other parts of the world. They developed distinctive cultures, invented languages, and, in what is now Mexico and Peru, established powerful empires whose built environments inspire the same wonder as the Great Wall of China, the temples at Angor Wat, Europe's cathedrals, Ethiopia's rock churches, and Egypt's pyramids. The Mayan creations of a written language, an astronomical system, and an indigenous and original mathematics are testament to the intellectual vitality of this crown jewel of the cultures of the Americas. Once settled populations were established, agricultural technologies developed that were just as vital, if not as striking, as the architectural ones. The agricultural gifts of these technologies include the tomato, the potato, tobacco, and corn. More quotidian, but as important, was the establishment in both the Americas of extensive trading networks and population centers. When Europeans finally arrived in the Americas a few hundred years ago, they did not discover an empty land begging for settlement and development.

What Europe perceived as a "New World" was populous (estimates run into tens of millions), widely settled, and home to a variety of established and accomplished civilizations. Although Polynesia and parts of what is now known as Micronesia were not yet occupied, the last major step in modern humans becoming a global species was completed with this settlement of North and South America.

The application of DNA sequencing and genomic analysis to the problem of human prehistory has brought deep and surprising insights into how we came to be who we are. Deployment of this powerful technology has shown that human populations have not been fixed and static genetic lines. Our population is not composed of races established in early prehistory and that remained genetically intact, unchanging over millennia until Europeans "discovered" the rest of the world and initiated the movement and mixing of populations. Migration out of, back into, through, and around areas has been the rule. From the beginning, our species has had a pattern of migration that is much more random and chaotic Brownian motion than inexorable linearity. DNA sequencing has shown that different human populations—in Africa, in the Eurasian landmass, and even in the Americas—are not pure long-standing genetic lines but dynamic genetic mixtures, ever changing. Some of the mixing took place long ago and involved many human populations that are extinct; some of it is in the recent past and is ongoing among today's populations.

WHY DID MODERN HUMANS SURVIVE
AND NEANDERTHALS PERISH?

When dark-skinned humans walked out of Africa, some took routes that, eventually, led them into what we now call Europe. They found humans already there. These archaic humans were the Neanderthals, and they had been in Europe more than 200,000 years before the arrival of these modern humans. Pale Neanderthals, some with red hair, met dusky immigrant populations of modern humans 45,000 to 50,000 years ago. This established population of humans who greeted our ancestors were skilled makers of tools, they made clothing from animal skins, and had learned to control and use fire. They buried their dead, perhaps on some occasions with flowers. They were creators of art. They were intelligent and had brains that were, on average, a bit larger than ours. As they met, our species and Neanderthals came to occupy the same areas. Based on the territoriality characteristic of higher primates, we can confidently assume that conflict and resource competition between bands of these two species were major themes of these interactions.

After confrontation, Neanderthal populations in Europe decreased and those of modern humans increased. Less than 10,000 years later, Neanderthals were extinct, but populations of modern humans continued an expansion that would overspread and become the dominant version of humans in Europe and the rest of the supercontinent of Eurasia. This expansion encountered, mated with, probably competed with, and perhaps also displaced Denisovan populations, eventually becoming the only human species on the planet.

Why was a smaller population of invading newcomers able to displace long-established ones? Since higher primates often use violence in their competition with members of their own and other bands, we would expect that our bellicose species must have fought Neanderthals bitterly, frequently, and lethally. On purely physical terms, in a violent confrontation, certainly Neanderthals would have been credible matches for us. Though shorter on average, they were more massively built and almost certainly stronger than our kind. There is little reason to assume modern humans had superior innate intelligence; recall that Neanderthals had a slight edge in brain size.

Regarding toolmaking, however, the African invaders arrived with a superior cultural package. The variety and sophistication of their tools and the techniques used for their manufacture were distinctly superior to those of the Neanderthals. Concerning hard power, these considerations are significant. As Stanford linguist and evolutionary theorist Marc Feldman has pointed out, the same techniques that produce better tools can produce better weapons. In armed conflict, modern humans probably would have had better weaponry than their more muscular competitors. In addition to hard power, modern humans may have also had the soft power advantage of a better social toolkit.

More effective forms of societal organization would not only have allowed them to field better fighting forces, but equally important, it could have made their gathering of raw materials and food more efficient and productive. If this was the case, an MBA might conclude that the invader's benefited from a stronger, more bountiful, and resilient economy than their brawny rivals. This would be in line with the evidence that the numbers of modern humans continued to increase during their co-occupation of Europe with the Neanderthals. Such population increases would benefit from the fruits of better organizational practices, enabling more complex social organization and further enabling even more efficient harvest and exploitation of the area's resources. More food would support an even larger population. This cycle of expand–exploit–expand would have been a virtuous one for the invaders and their descendants but a vicious one for the resident Neanderthals who would find themselves fighting bigger and bigger bands of their rivals and having to share an area's resources with more and more competing hunters and gatherers. Perhaps the economic soft power of modern humans was a more important cause of Neanderthal extinction than the hard power of brutal and remorseless interspecies pogroms. Moreover, they may have brought germs, new pathogens tolerated by them but not by Neanderthal hosts.

Competition—violent, economic, or microbial—was not the only avenue of interaction between these species. Comparative studies of the DNA sequences present in most of today's populations of modern humans reveal the presence of DNA from archaic human species. About 1.5 to 2 percent of the DNA in people of European ancestry come from Neanderthals, and in those of East Asian origin, both Neanderthal DNA (around 2 percent) and a small amount (around 0.2 percent) Denisovan DNA are found in the genome. The presence of Neanderthal and Denisovan DNA in the genomes of many modern humans demonstrates that these human species not only met but, in a version of love across genetic distance, mated and produced fertile offspring. The very low percentages of Neanderthal and Denisovan DNA in sub-Saharan Africans speaks to geographic distances enforcing a lack of genetic interaction between these populations. Even before the age of genome sequencing, anthropologists had speculated on the possibility of interbreeding between different species of humans. The advent of highly sensitive and precise methods for DNA sequencing have provided definitive evidence that interbreeding between humans and Neanderthals and between humans and Denisovans, too, was not only possible, it was widespread, frequent, and pervasive.

Was a significant factor in the extinction of the Neanderthals and Denisovans their assimilation into larger and more fecund populations of modern humans? Were these rivals in some sense just loved to death? While the signature of interspecies mating is incontrovertibly written in the genomes of many modern humans, the extent of hybridization was insufficient for interbreeding to absorb the Neanderthals and Denisovans. It seems that even though the barrier between the gene pools of these cousins and modern humans was somewhat porous and allowed a trickle of gene exchange, it was never massively breached. Therefore, we have to look elsewhere—perhaps back to various forms of interspecies competition such as those mentioned—for an explanation of why Neanderthals became extinct.

We finish this overview of the migration of our species in and out of Africa with notice of historical parallels that uncomfortably and ironically echo each other. The colonization of western Eurasia, what we now call Europe, by modern humans involved dark-skinned populations displacing indigenous pale-skinned populations of Neanderthals, probably by deploying a combination of better weapons and better social technologies. Tens of thousands of years later, we now see pale-skinned Europeans using better military and social technologies to colonize, subjugate, and, in some instances, nearly exterminate indigenous darker-skinned populations. We don't suggest a moral equivalency between the interspecific conflict involving *H. sapiens* and *H. neanderthalensis* and our own intraspecific conflicts. We do wonder if our dark-skinned forebears of tens of thousands of years ago were as mistaken in their perception of color as a signifier of innate superiority and privilege as some of our more recent light-skinned forebears have been.

Chapter Two

The Notion and Nature of Race

"There is more to being black than meets the eye."
—From Charles Gordone's play, *No Place to be Somebody*

"What is person 1's race?"
—Question on 2010 US Census form

In her book *Left to Tell,* Rwandan American author Immaculee Ilibagiza provides an example of the carnage that notions of race can inspire. Her story begins in 1994, a few days after Easter. On holiday from college, she was visiting her family at their home in the bucolic green and hilly Rwandan countryside. However, the serenity of the landscape was misleading; it was a time of anxiety. For several days, tensions had been simmering between the majority Hutu and the minority Tutsi communities. She knew that tension had turned to violence when she looked across a nearby river and saw a person being hacked to death with a machete. This was the beginning of an extended spasm of violence during which Tutsis were hunted down and slaughtered. Over the course of the next one hundred days or so, a genocide was carried out that left three of every four Tutsis, a total of 800,000, dead. Although it was conducted without the cold efficiency and industrial technology of the Holocaust, its impact on the targeted population was as terrifying and devastating.

This atrocity grew out of soil that had a long history of racial poisoning, for Rwandan society had long been stained by racial tension and myth. Historically, the Tutsi had been a cattle-raising people, and the Hutus were farmers and workers of the soil, which put them at an inferior position in the economic and social hierarchy of this part of East Africa. The colonial powers—first the Belgians and then the Germans—that occupied what is now Rwanda held the view that the Tutsi were taller, finer featured, more industrious, of Northeast African stock, and inherently superior to Hutu "Negroes" whom they saw as inferior and of sub-Saharan origin. The colonists introduced this virus of racial superiority among the Tutsi and took pains to nurture the infection, ensuring that it remained chronic. Since a divided indigenous population was easier to rule than a united one, it was in the interest of the European colonial overlords to encourage and fuel tensions between Tutsi and Hutu communities. Actually, and ironically, intermarriage had produced a significant blending of these populations, and such strict racial demarcations were far more imaginary than real.

Of course, there are examples of racially inspired atrocities much more familiar to Americans than the Rwandan genocide: the grim benchmark of racially motivated violence, the Holocaust, immediately springs to most minds. Efforts to exterminate entire populations echo from place to place and decade to decade. Thus, Hitler's application of racial categories to European Jewish populations echoed the efforts of Turkey earlier in the century to exterminate its Armenian population, in which a million and a half Armenians were killed.

We are keenly, and finally, after so many years, widely aware that our nation was founded and grew to its dominant position on land taken, largely by ethnic cleansing, from people European settlers considered different, less civilized, and inferior to themselves. Slavery and the atrocities that still follow in its wake are an ugly and stubbornly persistent stain on Old Glory. And we can't stop here.

Asian Americans insist we go on, citing the pervasive discrimination and sometimes the violence that Chinese immigrants experienced during the nineteenth century and the first half of the twentieth century. They would also remind us that the only race-based forcible confinement of a population since the abolition of slavery was the internment of Japanese Americans during World War II. Despite the comparable, perhaps more grave, threat posed by Germany and its Italian ally during those war years, there was no confinement or restriction of German or Italian Americans, both White populations. Although it is likely that race was a factor in the starkly contrasting treatment of Americans descended from our European adversaries as opposed to our Japanese ones, factors such as the much larger numbers of these populations and their degree of integration into the social fabric of the United States were perhaps determinative.

Continuing in a similar vein, added to a history including discrimination and disrespect, many Mexican Americans are deeply offended by a major American political party supporting or acquiescing in the intention to build a wall that would separate large portions of the United States from our neighbor, Mexico. This measure is claimed to be essential to secure the long border between our two countries. However, there is no clamor to build a wall to secure the much longer border between the United States and Canada, a nation with a largely White population.

IS RACE SOCIAL CONSTRUCTION OR BIOLOGY?

What is race? It means one thing to some, something else to others, and there are those who deny it has any meaning at all. Whatever race may or may not be, most Americans think they belong to one. According to the 2010 Census, 97.6 percent of people in the United States, nearly 299,736,465 individuals, claimed membership in racial categories such as White, Black, Asian American, Native American, and South Asian. To some, race is a biological term, a real and objective biological category, similar to genus and species, describing a subdivision of nature that exists without reference to social construction. However, there are others who view race as a purely arbitrary notion—a social construct created by particular cultures—that has no more biological reality or significance than the label Democrat or Republican. Whatever race is or is not, it has been a determinative factor in the history and culture of the United States.

Race stubbornly persists as a dominant factor in American culture and in other cultures of the world. Later, when we define race, an essential element of the definition will recognize that races are populations whose members are more likely to mate with those within their racial population than with those outside it. However, we will also make social construction an important and usually major part of what we recognize and define as race in humans. The definition we offer later in this chapter insists that human races have identifiable cultural attributes as well as characteristic frequencies of some inherited traits. Such a formulation exposes and captures some of the complexity the term *race* carries when applied to human populations. Race should be recognized as a marriage of both culture and biology that is always heavily freighted by culture and, when legitimate, has at least some biological correlates, too. Because of its accessory or prime role as instigator of so many atrocities, *race* is a troubling term and is burdened with a great deal of social baggage that, quite appropriately, requires a lot of explaining.

Some who worry about the acceptance of race as anything more than a social construction point to mischief caused by the naïve and less than critical acceptance of "race science." Race pseudoscience encourages racial stereotyping and easy coupling of superficial physical traits with character and behavior, as well as to the pernicious assumptions that these physical traits are indicators of the inherent moral and cultural superiority of some races over others. As already noted, such views allowed industrialized European societies to rationalize the taking of land, resources, and control of their

destinies from long-settled and established Asian, African, and Native
American ones.

In times closer to our own, a twisted race science was a major pillar
supporting the genocidal racial policies that intended extermination of whole
peoples. The crucial factor responsible for the introduction of such racially
engendered aberrations into the pattern of societies is not the perception of
difference but the assumption of racial superiority. Some societies have been
much too eager to embrace the notion that the lower technological accomplishments of other societies were inherently, and in important ways, genetically determined rather than being a consequence of alternative trajectories
explained by cultural differences, physical environments, and the vagaries of
history.

Those who urge denial of race as biological reality don't find it hard to
find examples, some egregious, of social or cultural definitions of race that
have been designed to serve less than benign purposes. There are many
instances where the definition and recognition of race have been designed to
mark individuals or groups for exclusion, exploitation, or even worse. These
definitions are indeed social constructs that have been created to serve juridi-
cal or political purposes rather than to explain or classify biological realities.
As examples, consider the following cultural fossils dug up from a bygone
era. During a dark period of German history, the Reich Citizenship Law, one
of the anti-Semitic Nuremburg Laws, defined membership in the Jewish
"race" as follows:

> A Jew is anyone who descended from at least three grandparents who were
> racially fully Jews. A Jew is also one who descended from two full Jewish
> parents if (a) he belonged to the Jewish religious community at the time this
> law was issued or he joined the community later, (b) he was married to a
> Jewish person at the time this law was issued or married one subsequently, (c)
> he is the offspring from marriage with a Jew contracted after the "Law for the
> Protection of German Blood and German Honor" became effective, (d) he is
> the offspring of an extramarital relationship with a Jew and was born out of
> wedlock.

In the United States, laws banning miscegenation, primarily marriages between Blacks and Whites, intended to protect the White race from pollution
by the Negro race, were widespread throughout the American South and in
many Border States. A couple of examples show that these laws relied on
legal definitions of "Negro" in terms of "blood," that were biologically dubi-
ous, often circular, and varied widely.

> A miscegenation statute of Missouri decreed it a felony for Whites to
> marry Negroes and that Negroes "are defined as persons having one-
> eighth part or more of Negro blood."

> An Arkansas statute stated that it was a misdemeanor for any White person to marry "any person who has in her or his veins any Negro blood whatsoever."

During the Nazi era, laws containing such peculiar and socially constructed definitions of race were created to facilitate the removal of Jews from all aspects of German life. The pursuit of this malevolent societal goal led to wild schemes for detection of Jewish identity, one of which involved using a pendulum as a sort of divining rod to detect individuals with "Jewish blood," a ridiculous strategy that would have been comical if the context had not been so tragic. Certainly here in the United States there has been a preoccupation with race and racial fractions, with some Whites now expressing great pride in the discovery of fractional Native American ancestry. However, a similar eagerness to learn of Black ancestry is not common. As illustrated by the fall of the protagonist in Sinclair Lewis's novel *Kingsblood Royal,* not too long ago such knowledge could transform lives in unwelcome ways. This tale tracks the fate of a White family whose social standing and ability to earn are put at crucial risk by the revelation of Black ancestry. Moreover, as was seen even a few years ago, the furor stirred up by the revelation that Thomas Jefferson is the founding father of two bloodlines, one White and the other Black, shows that who is defined as one race or another still raises eyebrows and can cause unease.

This history of past outrages, the witness of current depredations, and the reasonable expectation of future episodes of racially motivated mistreatment, even violence, have inspired some of those interested in bettering intergroup relations to attempt taking race off the table by denying the existence of human races. If there are no human races, there is no basis for racially inspired bigotry and the evil resulting from its practice. Furthermore, some feel that the concept of race, no matter how carefully formulated, is without biological reality or meaning and that it is only a social construct, like being a Republican or Democrat, a German or an Austrian. After all, there is no innate anatomical or genetic difference between Republicans or Democrats nor between Germans and Austrians. All that is necessary to convert one into the other is to change party affiliation or move an artificial line called a border. For those who take this view, race is just a label; there is no biology.

However, beyond assertion, is there really hard evidence that there can be more to a racial label than arbitrary social construction? Are all differences between groups designated as races politically and culturally constructed or is there, in some cases, also a deeper biological reality?

IS THERE ANY CONNECTION BETWEEN
A RACIAL LABEL AND BIOLOGY?

Imagine a thought experiment to determine if we could identify racial populations without resorting to visual cues. Suppose we assembled three hundred US citizens, a third of whom had identified themselves as racially Asian on the 2010 census form, another third who checked the African American box, and the remaining third who declared themselves White. Further, suppose that they agreed to go along with our little experiment and each group entered a room into which we could not see. Would it be possible to determine which room contained each of these groups?

Absolutely. We need only obtain a cheek swab from each of the people sequestered in their racially designated rooms, extract the DNA, and subject it to DNA sequencing. The results of this analysis would unequivocally tell which room contained the White, the Asian, and the African American populations. While it is true that these racial labels do have important elements of social construction and cultural context, without at least some degree of underlying biological reality, our thought experiment would have failed. In fact, something very much like our thought experiment has already been performed and published in the scientific literature.

Using an appropriately programmed computer, Dr. Neil Risch's group at Stanford University has analyzed DNA samples from a large group of individuals, the majority of whom self-identify as Asian, Black, or White. Using only data from the analysis of DNA, the computer sorted these individuals into three groups: 1, 2, and 3. A look at the record revealed that all of the individuals in Group 1 identified themselves as Asian, those in Group 2 declared themselves White, and those in Group 3 said they were African American. It is striking that the computer was able to use a purely biological criterion, a DNA sequence, to sort them into groups that matched the racial group they self-identified with an accuracy of 99 percent. This data makes it difficult to maintain that the notion of race has no biological reality. Two things are clear—there is more to race than meets the eye, and race, like beauty, is more than skin deep.

Although the Stanford results make it clear that members of the same race do have more of some hereditary components in common with each other than they do with other racial groups, it does not mean that all members of the same race are alike. As well as genetic diversity between racial groups, there is enormous diversity within them as well. It is not as though each race really did come from a single Adam and Eve, with every member tracing his or her ancestry back to a single pair of great-great-great-great-etc.-grandparents. The ancestry of all races is mixed, as though every race has had not one but multiple Adams and Eves, shared to a large extent but not completely by other races. Also, through interracial mating, racial populations continue to import and export genes, regularly introducing new and different sources of inheritance, shuffling and diversifying the genetic deck that comprises a racial gene pool that is and has been, to some degree, open. Members of a particular race differ so much from each other because they are each the offspring of different parents, each of whom traces a different lineage containing, to varying degrees, contributions from other breeding populations. Races are not uniform collections of individuals; they are inherently internally variable. A good cook would think them as much more like complex stews than homogeneous sauces.

Imagine how little meaning it would have regarding the overall nature of a good seafood stew to describe or specify in great detail the composition of a single clam or hunk of fish. What one is interested in is the overall composition—the steaming and complexly variegated whole. Similarly, because of their great internal diversity, one does not learn much about the nature of a race by studying just a single member. Because of the diversity within races, the concept of race has most meaning when applied to populations rather than to individuals. It is important to bear in mind that descriptions of the biology of races are statistical descriptions of populations reflecting averages for a group. Such descriptions are not based on any individual member of the race. Each member, while incorporating some of the ancestral and cultural features describing the race, has a unique genetic and cultural profile.

THOUGH SOCIALLY CONSTRUCTED RACE IS TO VARYING DEGREES BIOLOGICALLY ANCHORED

While it is true that some features of race are merely social constructs, certainly some of the features that the person on the street distinguishes as White, East Asian, or African American have a basis in biology and a reality independent of culture; the most biologically meaningful definitions of race flow from the recognition that races are breeding populations, groups whose members are more likely to mate within than outside the group. If a group persists in mating more within than outside the group, the composition of its gene pool will differ to some extent from that of other groups. The biology of race arises from the fact that breeding populations are gene pools, each with its own idiosyncratic makeup. While these gene pools may be distinctive in some ways, in human populations, the gene pools of breeding populations greatly overlap each other, and all have most genes in common with each other. As we will see later, to varying degrees, cultural factors, ancestry, and geography are responsible for the creation and maintenance of human races, recognizable as breeding populations where the likelihood of mating is greater inside the group than outside it.

Consider Whites and African Americans, two of the racial groups identified on US census forms. As we will see, although these populations have
greatly overlapping gene pools, they are not identical and can be distinguished. The population we label "White" traces ancestry back to Europe
while the population labeled "Black" is a European/sub-Saharan African
hybrid race created and maintained by social factors. It has a large ancestral
component that is African and a significant European contribution, as well.
Even though these two populations have shared the same broad geographical
area with each other and with other populations for almost 400 years, a
combination of social factors has made it more probable that members of
these groups mate within rather than outside their own population. The biological result of maintaining these breeding populations is apparent and manifest to the eye. The cultural results are just as apparent and have vastly more
impact.

RACIAL IDENTITY IS NOT
ELECTIVE FOR MOST PEOPLE

While biologists have something to contribute to an exploration of race in
human populations, Barack Obama's declaration of Black identity provides a
convincing example of the contribution of culture to meaningful notions of
race and racial identity. An unequivocal mixture of equal parts White and
African ancestry, what race is the forty-fourth president? His birth certificate
ignores the question, leaving it to him and the culture he inhabits to provide a
label. His choice settled the matter. On the 2010 census form, President
Obama checked the box labeled "Black, African-American or Negro," bypassing the opportunity to declare himself biracial by checking more than
one box. He married a woman who says she is Black, the Obamas say their
two daughters are Black, and President Obama ran as a Black politician in
both local and national elections. He has declared membership in a Black
church. He is Black/African American/Negro because he has chosen to socially construct himself as such, providing a clear illustration of the power
and importance of culture in understanding race.

However, ask yourself, would this decision have been considered a legitimate one if he had had no sub-Saharan African ancestry? Probably not, for
reasons explored shortly. For a time, in recognition of his empathy and
understanding of Black concerns, Bill Clinton was jokingly called "The First
Black President" by some in the Black community. However, it was never
understood as anything other than jest by anyone inside or outside of that
community, because President Clinton has never claimed to have any Black
African ancestry. Labels are cultural; ancestry is biological. The most meaningful racial labels usually incorporate both culture and ancestry. Both came

to be critical factors, important and determinative, in how Barack Obama saw himself, in how he was seen, and in the racial category he chose. However, returning to the census form, suppose he had decided to declare himself White, not Black or biracial. This choice would have been regarded as illegitimate by many, and in Black communities, he would have been viewed as confused or as a sort of traitor or possibly both. In general, something akin to the infamous "one drop" rule (*any* Black ancestry) would, in the minds of many, have disqualified a claim of membership in the White racial club.

Black communities contain many people of mixed ancestry who might be judged as some race other than Black if encountered on the streets. A few of these individuals have chosen to pass themselves as White, abandoning one race and joining another. These have usually been people with skin tones light enough to be decoded as White. This behavior, called "passing," is a choice that carries considerable risk. Until relatively recently, Blacks who were discovered to be passing as White were, at the very least, socially embarrassed and stripped of the White privilege they may have gained by their subterfuge.

Until recently, here in the United States, sanctions more severe than shame were very much on the menu. Since there were laws in many parts of the country forbidding: interracial marriage; sale or purchase of property covered by restrictive covenants forbidding sale to Blacks; voting; rental of a hotel room; or purchase of a meal in establishments that explicitly stated a policy of "Whites only" or coded the exclusionary policy with a "We reserve the right to refuse service to anyone" sign; employment reserved for Whites only; and so on; individuals found to be passing as White could be sued, arrested, jailed, or all of the above. And of course, as Coleman Silk, the Black-passing-as-White protagonist of Philip Roth's *Human Stain* learns, they lived under the threat of physical intimidation and injury. Look back to an earlier time and imagine the fury a White family might have turned on the Black man whose passing enabled his marriage to one of the family daughters.

Historically, many in the Black community have viewed "passing for White" as something like treason. There is the perception that individuals who pass are using an accident of biology—their skin color, features, and hair texture—to gain advantages denied to other members of the Black community. Just the reverse has been felt for those who, despite their capacity to pass, chose not to do so. Walter Francis White, the fair-haired, blue-eyed, and highly effective director of the National Association for the Advancement of Colored People (NAACP) during the early to mid-twentieth century, exemplified the kind of loyalty to the Black community admired in those who could have passed as White but instead chose to serve the interests of the community they were born into.

TALES OF IDENTITY CONFUSION AND CRISIS

The extraordinary golfer Tiger Woods, already placed among the great figures in the sport, is a person of mixed ancestry. During the time he dominated golf the way few have, his success inspired an interest in golf among many in the Black community, which was not a population who had paid a lot of attention to the sport before the era of Tiger Woods. Their attention had been captured by the ground-breaking achievement of athletes such as Althea Gibson, sisters Venus and Serena Williams, Arthur Ashe, Muhammad Ali, and Jackie Robinson. Now with the conspicuous success of Tiger Woods, many Blacks felt that he had proved that Blacks, too, could be outstanding golfers. However, Woods declined to identify as Black, and there was disappointment when, in recognition of his rainbow inheritance including Black, East Asian, Native American, and Caucasian forebearers, he declared himself "Cabalinasian," a neologism that recognizes all of these. Some of those who recall a time of less freedom to opt out of being regarded as Black enjoy a racial joke that tells the following story:

> Aliens use time travel and other technologies to visit Earth and kidnap Tiger Woods in order to discover the secret of his golfing talent. The abduction goes well, they pick him up in 2019, take him into their spaceship, sedate him and conduct the extensive probing they think is needed to determine the basis of his superiority. They learn what they wanted to know, wake him up and ask where on Earth he'd like to be returned. He says "just drop me off at the gate of the Augusta National Golf Club." They carry out his wishes, but make a timing mistake and drop him off in 1959 instead of 2019. The guards, otherwise decent fellows, but people of their time and place, challenge his attempt to enter the Club, stating: "No Negroes allowed." He replies that he is Cabalinasian, not Negro. The guards don't understand the term and one of them asks him to write it down. He does so and hands it over. The guard takes a look, squints skeptically, turns to his colleagues and asks, "Ever see nigger spelled like that?"

This little joke takes the liberty of using an often taboo term that liberally populates many hip-hop songs and the routines of Black standup comics. It makes the point that although Woods does not embrace Black identity, many Whites and others will assign it to him anyway. President Obama used the freedom and opportunity gained through the sacrifice and effort of so many who fought, and some who died, for racial justice to choose an identity that would cast him as an inspiring role model. In contrast, Tiger Woods used the liberty given him by those struggles and sacrifices to define his racial identity as something other than Black. However, as we shall see, not everyone has been allowed to assign themselves to whatever racial category they wish, though with varying degrees of success, people have tried and some still do.

There is the strange case of Clarence King, a blond, blue-eyed Brahmin from Newport, Rhode Island. King, a Yale graduate and distinguished geologist, was the first director of the US Geological Survey. He wrote *Systemic Geology*, an authoritative work on the geology of the American West that was recognized as a comprehensive and definitive work. Mount King in Utah is named in his honor, as is Clarence King Lake in California's Mount Shasta wilderness. In her book, *Passing Strange: A Gilded Age Tale of Love and Deception Across the Color Line*, Martha Sandweiss tells how King, assuming the name Clarence Todd, passed himself off as a pullman porter and met, courted, and married Ada Copeland, a Black woman and ex-slave. Five children, four of whom survived, were born during their decades-long, common-law marriage. Through their many years together, his absences from the home were explained as work-related since it was easy to claim that his job as a pullman porter required extended periods of travel. During these periods of absence, he returned to life as Clarence King, a White man and distinguished geologist.

After his death in 1901, his dual identity became known when his Black wife, Ada Todd (nee Copeland), attempted to claim a trust fund that her husband, the racial and social chameleon, King, told her had been set aside for her and the children. A segment of the press, then as now, greedy for the sensational, ate the story up and vomited out the ugly and lurid headline: "Mammy bares life as wife of scientist." Although stories of passing from Black to White were not at all rare, there was surprise and amazement at the revelation of the strange passing of this descendant of signers of the Magna Carta and member of the American elite who had stepped across the color line, leaving the pomp and bright privilege of the White side to inhabit the different, difficult, and darker one of African Americans.

The weird saga of Rachel Dolezal, a White woman who passed as Black, is a reminder that a hundred years after Clarence King's episodic transracial migrations, racial choice remains limited, even for White people. Born of White parents who state their ancestry as Czech and German, she was raised in Montana and received a college education at Belhaven University in Jackson, Mississippi. Dolezal then applied for admission to graduate school at Howard University, America's most distinguished historically Black university, where she earned a Master of Fine Arts.

For a complex set of reasons that might include being reared by White parents who adopted Black children and, perhaps because of experiences at Howard, Dolezal increasingly came to identify herself as Black, discarding the racial label matching her genetic makeup. During the course of her transracial journey, she married a Black man and made cosmetic changes that included darkening her skin and using hair weaves and dye to transform her straight blond hair to a curly corkscrew cascade of a darker hue. Eventually, she moved to the Northwest and, for a time, was an instructor at Eastern

Washington University, teaching courses in the area of African American and Africana Studies. In Spokane, Washington, she identified herself as Black and got involved in Black community activities. She even became president of the local chapter of the NAACP. Reports are that she did a good job as president.

Her passing as Black came to an abrupt and dramatic end when her White parents came forward with a public declaration that she was White and had no Black ancestry. Public reaction was immediate, widespread, and almost uniformly negative and censorious. "Confused," was the most charitable of the many terms of censure used to label her. Dolezal was accused of taking advantage of this masquerade to advance her career and status. Some critics accused her of exercising White privilege; others said she lacked a sincere dedication to the advancement of the interests of Black people. Passing, whatever the motivation, always employs subterfuge and is, in fact, lying. As a consequence of her misrepresentations, she lost her academic job, was forced to resign her position as president of the Spokane NAACP chapter, and continues to be a target of anger, disdain, and vilification. Still, in spite of the price she has paid, Rachel Dolezal continues to struggle to maintain a transracial Black identity. It seems that in her mind, she is not a White person pretending to be Black but a transracial Black, a once White person now transformed, she insists, into something different and Black. Ignoring her genetic makeup, she persists in her decision to socially construct herself as a Black person, going so far as legally changing her name to Nkechi Amare Diallo, a name one might hear in Nigeria. As of now, her claim to be Black, benignly misguided, but tenaciously and probably sincerely held, lacks ancestral cred and, therefore, is not widely accepted.

The routinely outrageous antics of our times make it hard to invent scenarios more absurd than some of what crawls out of our Twitter feeds. However, in *Get Out*, Jordan Peele's imaginative and hilariously creepy film depicting the ultimate in transracial migration, he offers scenarios of transracial conversion even more bizarre than what we read in the papers. In this movie, carrying a faint odor of that old warhorse *Invasion of the Body Snatchers*, aging Whites, gripped by fears of failing physical competence, concoct a scheme involving romantic entrapment, hypnosis, and the migration of consciousness to take over the bodies of Blacks.

At the physical level, the passing couldn't be more successful, but the changelings fail to display the souls of Black folk and so would not have fooled W. E. B. Du Bois or most other Blacks either. Right up until the end, the Black body of the film's initially credulous and increasingly terrified hero appears likely to become the reluctant host of a White consciousness. Then, in a hilarious *deus ex machina*, the bumbling efforts of an Inspector Clouseau–like TSA official, surely among the least likely sources of salvation, causes these macabre attempts at transracial passing to run aground.

WHY ARE THERE RACES, AND WHY ARE WE NOT SURPRISED THAT THEY HAVE BIOLOGICAL CHARACTERISTICS?

Why are there races? Something like what we call race was the inevitable consequence of migration. As they traveled and settled, early human populations found themselves in different environments and responded to the challenges posed by these new circumstances. These responses were always cultural and sometimes biological. The cultural responses covered a spectrum that included variations in social organization—populations of nomadic bands here; some populations eventually beginning the practice of settled agriculture there; various inventions of new tools and weapons; different populations perceiving different revelations of effective strategies for worshiping and manipulating the supernatural forces they thought responsible for their existence and the way things work; and, of course, linguistic diversification and evolution. Sometimes, the challenges of new environments selected characteristics that were biological, increasing the prevalence of traits that favored the survival of the population. As we will see, the evolutions of variations in skin color, disease resistance, body structure, physiology, and biochemistry were among the many biological keys to the success of humans as a global species. Some of these are associated with, but do not define, race, a category that involves much more than a single genetic or cultural feature.

From an origin somewhere in Northeastern Africa, we began the great migration that populated all of Africa and continued with the dispersal of humans out of Africa into Eurasia, Australia, and, finally, the Western Hemisphere. This migration scattered populations, eventually isolating them geographically from each other for long periods, making significant intermixing with distant populations a practical impossibility. This geographic dispersal created geographically separated ancestral human population clusters. Both in Africa and as they traveled outward, populations interbred with peoples they encountered, creating new diversity and new populations. The obvious differences in the appearance of individuals from these widely dispersed populations inspired Carl Linnaeus to create a simple system of racial classification.

AN ERA OF "SCIENTIFIC RACISM"

Everybody who has taken a biology class knows that living things are classified by genus and species—dogs are *Canis familiaris*, cats are *Felis catus,* and the apple that got Adam, Eve, and the rest of us into so much trouble is *Malus domestica.* We owe this system to Carl Linnaeus, an eighteenth-century Swedish scientist appropriately recognized as one of biology's great figures. A knowledgeable and widely curious man, Linnaeus was struck by what he saw as the great diversity of human beings that European voyagers were, at least in their view, discovering. Struck by external differences, particularly in skin color, between his Northern European colleagues and the varieties of individuals brought from East Asia, sub-Saharan Africa, and indigenous populations of North America, he was heavily influenced by geographical considerations to group humans into four races:

European: White, muscular, acute and inventive, gentle, sanguine, governed by laws

American: Red, erect, happy and free, choleric, obstinate, and governed by custom

Asian: Sallow, stiff, haughty and avaricious, melancholy, and governed by opinion

African: Black, relaxed and indolent, phlegmatic, and governed by caprice

These characteristics were declared constitutional, and these physical and behavioral features were assumed to be biological properties inherent in the natures of all members of these groups, traits passed down from parents to offspring. Later, toward the end of the eighteenth century, Johann Blumenbach added a fifth geographical race, the Malayan, or Brown race, to include Southeast Asians and Pacific Islanders.

Incidentally, it was Blumenbach, inspired by his belief that the southern Caucasus region produced the world's most beautiful people, who gave the name Caucasian to White peoples. Though certain as Linnaeus was of inherent disparities in racial abilities and behaviors, Blumenbach on the other hand, believed that non-White peoples had the innate capacities to perform in intellectual and cultural arenas at the same level as Whites. This was a strikingly uncommon view at the time. He even went so far as to sharply criticize colleagues who subscribed to the then-current view that these "inferior" races were incapable of doing so. The notion of race and the short list of races proposed by Linnaeus, Blumenbach, and others were widely embraced by Western science.

At its apogee, elaborations and extensions of racial science commanded the allegiance of some of the late-nineteenth and early twentieth century's distinguished thinkers. Among the devotees, we find Frances Galton, Darwin's brilliant cousin, who is often called the "Father of Eugenics," a neologism for "good breeding." An upper-class Englishman of extraordinary gifts, he wrote widely on the improvement of human potential by encouraging mating among the more fit and discouraging it among the less fit, extending his ideas beyond individuals to populations. Along these lines, he wrote a proposal for the improvement of Africa by encouraging Chinese, whom he deemed more fit, to colonize and outbreed the resident African populations whom he considered less fit.

In the United States, there was Louis Agassiz, founder of Harvard's Museum of Comparative Zoology and a leading intellectual figure of nineteenth-century America, who theorized that Blacks had separate evolutionary origins from Whites. In the early twentieth century, Charles Davenport, a leader in the eugenics movement and then director of the Cold Spring Harbor Laboratory, was an energetic crusader against miscegenation (a term coined to refer to interracial mating) involving White and Black populations. Davenport was an eager advocate of anti-miscegenation statutes, such as Virginia's Integrity Act of 1924, which barred marriage between a White person and anyone with any trace ancestry other than Caucasian.

At that time, "scientific racism," with its notion that physical features are indicators of intellectual and cultural capacity, was very much in the air in the United States, and this was reflected not only in legislation but, as mentioned below, in literature, too. The eugenics movement even included some of the more celebrated members of America's liberal tradition, for example, the energetic Theodore Roosevelt, a liberal leader who used his gifts of persuasion and still-celebrated presidency to force otherwise highly progressive legislation into law. Moreover, Margaret Sanger, the founder of Planned Parenthood, among the more ardent eugenicists, whom James Watson and Andrew Berry quote in their book, *DNA*, as saying: "More children from the fit, less from the unfit—that is the chief issue of birth control." Notably, Watson, who, with Francis Crick, discovered the structure of DNA, is highly critical of naïve eugenics and he pokes fun at the notion in the pages of *DNA*. This is in contrast with his Lear-like transformation of later years that have seen him lurch into a position that departs from the spirit of his earlier clear-eyed dismissal of the eugenics movement.

In the first half of the twentieth century, tracts by enthusiastic but less benevolent eugenics propagandists, notably Madison Grant's *The Passing of the Great Race* and Lothrop Stoddard's *The Rising Tide of Color Against White World-Supremacy,* sought to raise alarms about the challenge posed to the White race by colored races. Novels of the time reflect some of these concerns, some with derision. For instance, in *The Great Gatsby,* F. Scott Fitzgerald's masterpiece, the privileged and brutish Tom Buchanan is satirized as a shallow intellectual pretender who spouts a bigoted and superficial mishmash of themes from these deeply flawed, but in their time, widely read and highly influential books by Grant and Stoddard.

A THOUGHT EXPERIMENT THAT GENERATES A RACE

So what is a race? Another thought experiment will be helpful in conceptual-
izing how a race might arise, and it will allow an appreciation of the definition of race soon to be offered. Suppose that in the not-too-distant future, in
yet another migration, humans establish an increasingly self-sustaining colony on Mars. The founding colonists are all in good health, are of reproductive
age, and were recruited from a dozen countries scattered around the world,
including China, Nigeria, Norway, Egypt, India, Mexico, the United States,
New Guinea, and a few others. Although there are contacts between the
migrants and home—trade missions and the occasional trip back home—the
time required to transit the 47 million miles back to Earth and the expense of
the trip means that these interchanges will be minimal. Most of the colonists
and their offspring will spend their lifetimes knowing a great deal about
Mother Earth but never setting foot on her. Though not forbidden, given
these practical realities, migration from colony to Earth or Earth to the colony is likely to be only a trickle, perhaps only a dozen or so individuals per
year. The initial colony of 300 establishes itself and after a thousand years,
time enough for about fifty generations, has grown to a population of 10,000.

They would share cultural experiences of technology, lifestyle, aesthetics,
and, perhaps for some, religion found on Mars. While some of these would
be similar to those practiced on Earth, many would be distinctive, inspired,
shaped, and developed in an environment that afforded experiences and in-
spirations that were uniquely Martian. Perhaps a Martian language, certainly
a Martian dialect and vocabulary, would develop. After a few centuries, a
meeting between a group of Martians and people from Earth would expose
any number of cultural differences, some subtle and others, such as language,
dietary preferences, and religious practices, that are quite different. Although
occasional mating between Martians and the residents of Earth would take
place, most of the mating of Martians would be with other Martians. They
would constitute what a geneticist would recognize as a breeding population.

Comparing a photo of the original group of 300 colonists with a randomly
chosen group of 300 of the fiftieth generation of their descendants would
certainly show differences. In all likelihood, the rainbow-like diversity of the
founding mothers and fathers would be replaced by something different and
perhaps more uniform in their great-great-great-etc.-grandchildren. DNA sequencing, however, would easily pick up traces of the DNA sequences
brought to Mars by the founding population, but time and the many generations of matings in the new environment would have produced a population
with frequencies of some DNA sequences distinct from those of the founding
population. Eventually, the Martian breeding population would have a gene
pool different from the one brought in by the founding population. They
would have a different culture, too. They would be a different race.

This little allegory illustrates several features of race. Races are breeding populations in which matings are more likely to take place within than outside the population. Breeding populations can be created and maintained by geographic separation. The great migration that resulted in the global distribution of our species geographically isolated human populations from each other. For most of human history, populations in Africa could not mate with distant ones in Asia; East Asian populations could not mate with those in Europe; North American populations could not mate with those in Africa; and so on. Even in today's world, interconnected in so many ways, distance is still a significant barrier to a free exchange of genes. In addition to geographic barriers, cultural ones can also be important and sometimes paramount. Even within the same nation, cultural barriers such as law, tradition, or religion may discourage or even forbid mating between certain groups. Consequently, cultural as well as geographical factors can create and maintain breeding populations, groups in which more matings are within the group than outside the group.

Maintained over many generations, breeding populations constitute gene pools that are slightly different from the gene pools of other such populations. The advance of technology has made it clear that DNA sequencing can demonstrate these genetic differences. Because races are, in a loose sense, breeding populations, it is possible to identify a cohort of genes that have frequencies that allow distinction among these populations. At most, these genes constitute a very tiny fraction of the genome. Some of these genes specify differences in appearance—skin color, eye color, whether eyes are round or almond shaped, hair color and texture, and so on. However, most encode differences that are internal and invisible, requiring sophisticated clinical or biochemical tests for their detection. Many members of a particular racial population will self-identify as members of the racial group. However, the assumption or denial of membership in a particular racial group may be tightly controlled by cultural construction.

FINALLY, A DEFINITION OF RACE

We will use this definition of race, which includes both culture and biology: *A race is a breeding population delineated by culture that has characteristic or distinctive frequencies of some genes.* As we proceed, culture will emerge as the most important factor in determining most of the differences observed between races. However, ancestry—genetics—is an essential (and difficult to ignore) feature of race. In reality, races are not individuals; they are populations of individuals who are often quite diverse.

Consequently, even within a race, there is usually cultural and biological variation among individuals. Therefore, it is not surprising that individuals who claim the same racial identity on a census form may not look that much alike. Individual members of a racial group may differ in some of the cultural features associated with the racial group. For example, one person identifying as Han, the dominant Chinese ethnic group, might speak Mandarin, but English might be the principal language of someone else with an equally legitimate claim to be Han Chinese. Despite the differences in language, a purely cultural characteristic, both are willing to accept the other as Chinese. Similar latitude might be allowed for biological traits associated with particular races. Though differing in such genetically determined features as skin color and hair texture, both Barack Obama and Martin Luther King Jr. identify as Black and would have checked the same box on a 2010 census form. However, if either Obama or King claimed to be Han Chinese, that claim would have been generally rejected. Neither is culturally Chinese, and neither would claim even a little Han Chinese ancestry. Even if they spoke perfect Mandarin and adopted Confucian values, in the eyes of many Chinese, their lack of Han Chinese ancestry would disqualify them. Returning to our two previously mentioned Han Chinese people, even if they had attended a historically Black college, socially integrated into the fabric of campus life, and married twin Black sisters, because they lacked a credible claim of at least some Black ancestry, neither would be viewed as Black. While in no way denying the importance of culture, these examples illustrate the role played by ancestry in how race is viewed and determined in the United States and some other places, too.

WHAT DOES ONE HAVE TO BE TO LAY LEGAL CLAIM TO BEING NATIVE AMERICAN?

Who is a Native American? This question, a part of the American Experience, provides a thought-provoking opportunity to deepen our appreciation of how intertwined with culture, regardless of genetic ancestry, our recognition and assignment of identity can be. Indian identity again became a prominent part of the national conversation when Senator Elizabeth Warren, a devoted consumer advocate and highly accomplished former Harvard Law School professor, expressed pride in her Native American ancestry.

Warren, a blonde and blue-eyed White woman, based her belief in an American Indian ancestry on family stories handed down over generations. An Oklahoma native, she grew up believing that she had a great-great-great-grandmother who was Cherokee and assumed herself 1/32 Cherokee. Her recital of this distant connection with the Cherokee ignited a storm of controversy over whether she had any such ancestry, and some speculated that her claims were a factor in her appointment as a full professor at Harvard Law School. Donald Trump, while running for president of the United States in

2016, branded her "Pocahontas," a derisive taunt dragging the odor of a racial slur. Setting aside the distraction of President Trump's crude derisions, ignoring the fact that she did not claim to be a Cherokee, and failing to acknowledge Harvard's statement, clear and unqualified, that she was not appointed under one or another affirmative action program, if she had claimed to be a Cherokee, she would have had little to no chance of being acknowledged as one. She would not even be accepted as American Indian. Not because of the blue eyes, the blonde hair, or the white skin. There are blue-eyed, blonde Native Americans with very white skin and American Indians with gray eyes and red hair, too.

Senator Warren has not claimed to be Cherokee but has made reference to family stories that the family tree contains a Cherokee branch. Does this qualify her as Cherokee? To become a Cherokee, Warren would have to demonstrate that she has an ancestor who was documented as being Cherokee. Although an analysis of her DNA shows distant Native American ancestry, she has not used this to claim that she is a member of the Cherokee tribe of Native Americans.

Actually, it would be pointless to do so. She could not secure official recognition of tribal membership by simply offering DNA evidence, since DNA, by itself, does not provide an acceptable demonstration of Cherokee identity. She would have to be able to name an ancestor who met certain specific criteria. This is because the Cherokee Nation has decreed that only the descendants of individuals listed on the *Dawes Rolls of Citizens of the Cherokee Nation* are eligible for official designation as Cherokee. The *Dawes Rolls* are lists compiled between 1899 and 1906 of those living in the Cherokee Nation when they were forced off the land in what would become Oklahoma. The *Dawes Rolls* were used to allocate land to members of the Cherokee Nation. Furthermore, Warren would have to use this information to obtain a Certificate of Degree of Indian Blood (CDIB) issued by the Bureau of Indian Affairs (BIA). This certificate would certify her fraction of Native American ancestry.

In general, a person is an Indian if one of the Indian tribes recognized by the US federal government says you are an Indian. If no tribe certifies you as such, you are not officially an American Indian. There are more than 560 federally recognized tribes. This makes a long list containing familiar names such as Apache, Blackfeet, and Cheyenne along with many lesser-known tribes such as Shinnecock, Tejon, and Utu Utu Gwaitu Paiute that are unfamiliar to most Americans. Qualifications for membership vary among tribes, and tribal membership is somewhat like citizenship in a discrete political unit. This is an outcome of the recognition by the US government that Indian nations and tribes have a measure of sovereignty, an acknowledgment and gesture of redress for the forcible and massive appropriation of Indian lands and dissolution of Indian societies during the eighteenth and nineteenth centuries. Recognition as an Indian is a political matter, socially constructed and resolved by whatever ancestral qualifications an Indian tribe chooses to impose. As of this writing, Senator Warren has made no attempt to join an Indian tribe or to obtain a CDIB.

As explored earlier, in general, there are constraints placed on the freedom to elect a racial identity. While the terms *race* and *ethnicity* are sometimes used interchangeably, the constraints on changing ethnic identity are generally less rigid than those governing changing racial identity. Many years ago, Sammy Davis Jr., a noted entertainer and Black American, converted to Judaism. More recently, Ivanka Trump, daughter of President Trump, renounced her Presbyterian faith and converted to Judaism, joining the community of Orthodox Jews in a widely publicized change of ethnicity. However, as we have seen, changing races can be quite difficult. Of equal parts Black and White ancestry, President Obama encountered little resistance to his choice to live his life as a Black person, but here in the United States, with his known and visually apparent degree of Black ancestry, he would have had trouble assuming a White racial identity. As for Whites attempting to transition to Black, Rachel Dolezal's failed transracial journey and the strange and bizarre case of Clarence King are instructive. To paraphrase the line from *No Place to be Somebody*, Charles Gordone's Pulitzer Prize–winning play quoted at the beginning of this chapter, there really is more to race than meets the eye. Race *always* involves social construction and becomes a very slippery notion when there is no ancestry to provide a biological anchor.

RACES ARE POPULATIONS, NOT INDIVIDUALS

We close this chapter with the working definition of a race as a breeding population delineated by culture that has characteristic frequencies of inherited traits. Because there can be so much variation among individuals within a racial group, the word *race* is most meaningful when it is applied to populations rather than to individuals. Individual members of racial populations may vary greatly, perhaps in extreme cases containing individuals whose individual differences exceed the average difference between racial groups. For example, in the United States, as the racial label states, the White population is lighter skinned than the Black population. However, every now and then, one might encounter a Black person with a skin tone lighter than a particular White person. Some will say that this, and observations like it, provide a conspicuous rejection of the reality of race as anything biological. After all, if one easily observes differences among individuals within races that exceed the average differences between those races, then the concept of race has little biological meaning.

We say this does not weaken the notion that races can differ biologically and offer this reminder of everyday observations to help make the point. On reflection, it is apparent that some traits may vary more within some families than they do between families. Imagine a family, the Exemplars, in which parents of medium height have four children. Two of these are of medium height: one is barely five-foot-two, and the other is six-foot, exceeding the height of all the Exemplars and taller than his diminutive brother by almost a foot. Although members of this family vary more widely in height than the average differences between unrelated families, we have no difficulty granting that all of the Exemplars belong to the same family (especially if paternity is validated by DNA tests). Despite the intrafamily variation, it would not occur to us to deny that this family is biologically different and distinct from other families.

However, to avoid the impression that race means more than it does, let's examine one more thing. Human races, populations delineated by culture with characteristic frequencies of certain genes, are made up of individuals. Typically, members of a race, like the members of families, are diverse. Consequently, knowledge of racial identity tells us little about each of the individuals who identify as members of this racial group or that one. The statistical averages that are used as criteria for delineating and describing a racial group apply to the population, not to particular members of the group. Like the idea of "Average American," races—White, Asian, Black, and so forth—are statistical abstractions. Knowing that the average income for Americans was $34,489 in 2017 does not tell what *your* income was in 2017. Knowing that an individual identifies as a member of a particular race is not a reliable description of that particular person's appearance. More importantly, an individual's race is certainly not a personality profile, and it is not an assay of a particular ability or set of abilities. Consequently, knowledge of a person's racial identity will not allow determination of any of these qualities. The content of a person's character cannot be determined from the color of their skin. To know that, we have to know them.

Having explored the nature of race and found some reasons to regard it as something more than just social construction, we will use the next few chapters to examine the implications of race for health and medicine, and then we will wonder if race predicts athletic performance or is a determinant of mental ability. All of these are interesting questions. There is a large body of reliable data that will surely convince one that race impacts physiology and medicine. Questions about race and sports or race and mental ability ask whether race is a determinant of human potential or performance. These issues turn out to be very difficult to answer.

Chapter Three

Human Diversity

"The colors of the rainbow so pretty in the sky
And also on the faces of people passing by . . ."
—From *Somewhere over the Rainbow* by E. Y. Harburg and Harold Arlen

Get off an airplane in New York or London and look around. The riot of skin colors, hair colors, hair textures, and body types are outward indicators of human diversity. The fingerprint-based touch ID on our cell phones and the increasing use of biometric identification in airports are effective because each of us differs from the other in anatomy, physiology, cell biology, and ultimately, in our DNA. Like snowflakes, no two humans are exactly alike. It is now widely recognized that even our companion populations of resident microbes, our microbiomes, are unique to each of us, different from those of our neighbors. While biological diversity is apparent and unequivocal, we cannot understand, accept, and constructively live with human difference without awareness that diversity isn't just biological.

Our most important differences are cultural—how we learn, how we build, what we eat, how we organize communities, how we worship, how we raise children, who we marry, and who we don't. Many of these cultural differences have been better and more appropriately studied by social scientists than biologists. Nevertheless, there are biological differences between humans and between human populations, too. It would be unrealistic to ignore the reality of these differences, some of which are viewed through the cloudy lens of race. Although skin color is often, and incorrectly, used in isolation as a racial identifier, it and many other biological traits differ in frequency among human populations. Carefully selected ensembles of these biological markers can distinguish among populations belonging to the socially constructed categories we call races.

THE DEPTH AND VARIETY OF DIVERSITY

To illustrate this point, let's again ask 300 US citizens to participate in a thought experiment. Assume that 100 of these self-identify as White, 100 as Black, and 100 as East Asian. If they assembled into their respective groups, with just a glance, we could easily tell which group is East Asian, which Black, and which White. Dark skin color, hair texture, and other features on many of the faces would identify the Black group. Light skin color, a spectrum of hair colors ranging from black to very light, and something about the eyes, in many cases featuring pale irises surrounding darker pupils, and other facial features would distinguish the Whites. Straight black hair, light skin different in tone from the Whites, in many cases eye cast, and other facial features would identify the East Asian hundred.

But what success would we have if we could not make a visual inspection because each group had entered its own comfortable, but windowless, room. Just as in the previous chapter, the question is whether, without resort to appearances, we can tell which room contained which racial group. But this time assume we get no help from the DNA analysis that provided unequivocal and reliable answers in our previous thought experiment. Even without resort to DNA, it can be done if each of the occupants in each of the three rooms can be persuaded to do just three things: surrender a drop of blood, provide a sample of ear wax, and, over the course of half an hour, drink a quart of milk on an empty stomach.

Here are the results from each of the rooms:

- Room 1—Blood typing shows a higher frequency of individuals with blood type AB than either room 2 or room 3. Almost all of the ear wax samples are dry and crumbly. About half an hour after consuming the quart of milk, most (>90%) people in this room reported gastric upset and flatulence.
- Room 2—A higher frequency of individuals with type O blood than either room 1 or room 3; most of the samples of ear wax have a sticky and adhesive consistency. About half an hour after consuming a quart of milk, around 75% of the people in this room reported gastric upset and flatulence.
- Room 3—A higher frequency of blood type A and lower frequency of individuals typing Rh negative than either room 1 or 2; most of the ear wax samples have a sticky and adhesive consistency. The majority of people in this room reported no ill effects from milk consumption.

Just this data allows us to conclude with certainty that room 1 houses the East Asian population, room 2 contains the Blacks, and the White population is in room 3. So even without the power of DNA analysis, these traits, all of which are genetically determined and are displayed by populations identifying as one of these broad racial groups as opposed to the others, allow populations of Whites, East Asians, and African American Blacks to be distinguished from each other. But notice how careful we have been to couch this thought experiment in terms of populations, rather than individuals, citing statistical differences among populations rather than claiming the uniform presence of a trait in all members of one group and its total absence from another. In general, people who identify as Black or say they are White differ in skin tone and, here in the United States, some Blacks are not black and some Whites are not white. Some East Asian families will not have a single person who is AB blood type. After drinking a lot of milk, some Whites would experience gastric upset, the majority of East Asians and Blacks would experience some level of gastric disturbance, but some in each group might drink their quart of milk, happily experiencing no intestinal problems at all. While blood type and lactose tolerance have health and medical implications, many of the differences in groups designated as "races," though real and easily determined, have no known physiological implications. The majority of these differences are of interest only to students of human variation. What all of these traits have in common is that they are genetically determined and therefore stand-ins for DNA sequences. If we sequenced the DNAs determining these traits, we would validate the observed frequencies of these indicator traits at the level of the genes (DNA).

As indicated, the traits described above are just a few of the many genetically determined traits that differ in frequency across racial groups. Rather than sharp discontinuity—a trait being found in one group but not at all in others—what we usually see is considerable overlap between races. Although almond-shaped eyes are thought of as typical of East Asians, some Caucasian eyes are almond shaped and there are East Asians with round ones. While constellations of traits can be distinctive, associated with this racial group but not with that one, it is essential to emphasize that an individual trait is rarely unique to one racial group, but is usually seen in more than one, often in many. For example, dark skin, often regarded as a definitive racial trait, is actually widely distributed among many racial groups and, by itself, is not a distinctive racial trait. Although a hallmark of the populations of Black Africa and their derivative New World populations of African Americans, this trait is shared by many non-African populations. Think of Australian Aborigines, some populations of the Indian subcontinent, and Pacific Islands groups at one time clumsily lumped under the term "Negrito" or "Melanesian." It is also worth repeating that the vast majority of traits we associate with race have little influence on health and performance. It is difficult to definitively establish an evolutionary role for many of the traits we associate with race. A majority of race-associated traits are just passenger traits, along for the ride, but have little positive or negative influence on evolutionary fitness and selection.

Certainly, ear wax is one of those traits it is hard to believe played a determinative survival role. Most people get along very well without the cringe-inducing knowledge that ear wax comes in two forms, sticky and adhesive or dry and crumbly, an observation first reported by Dr. Matsunaga and his colleagues. Perhaps this finding is not the most celebrated of discoveries by this investigator, the observation that East Asian populations almost uniformly produce the dry, crumbly form of ear wax while the majority of Whites, Black Africans, and African Americans produce ear wax with a sticky and adhesive consistency, did provide stamp collectors of racial differences with yet another population-associated trait.

Interestingly, Dr. Matsunaga suggested that some of the components of the sticky form of ear wax are organic compounds that are also present in sweat and may undergo subsequent chemical changes that produce an odor that some Japanese and other East Asians find unpleasant. It has been speculated that this may be partially responsible for the perception in some East Asian societies that Westerners have a distinctive body odor that does not afflict their fellow East Asians. If so, this would provide a biochemical basis for the unintentional, but real, offense to East Asian noses, which some Westerners, despite a fetish for frequent bathing, may deliver. But ear wax consistency is not regarded as a signature indicator of race. Skin color is. So it is worth a careful look.

A LONGER LOOK AT SKIN COLOR

Skin color, and eye color as well, are determined by the kind, level, and distribution of melanin, a pigment. There are two forms of melanin, orange-red pheomelanin and brownish-black eumelanin. Pheomelanin gives the red color to lips and is the red in red hair. Eumelanin determines skin color and, with the exception of red hair, whether hair color is black, brown, or blond. Although many genes collaborate to determine skin color, some make greater contributions than others. Variants of the genes KITLG, MC1R, SLCA24A5, SLC245A2, TYR—acronyms for exotic and unfamiliar names like Kit ligand gene, melanocortin 1 receptor, solute carrier family 24 member 5, solute carrier family 45 member 2, and tyrosinase, respectively—are among the most important and familiar to researchers who study the molecular genetics of skin color and its variation. Like skin color, eye color—blue, brown, gray, green, or hazel—is also determined by many genes, at least sixteen of which have been identified. The most important of these are the genes OCA2 and MC1R, merciful acronyms for oculocutaneous albinism II and melanocortin 1 receptor, respectively. The color of the eye is determined by the interaction of the amount and distribution of melanin in the iris and the scattering of light by the turbidity of the iris. The amount and distribution of melanin is the major determinant of the color of the light reflected from the eye. Just as there is no blue dome in the sky to give it a blue color, there is no blue pigment in blue eyes. We see the sky as blue because the physics of light scattering are such that the blue portion of the visible light spectrum is most effectively scattered. Something similar is going on in eyes that appear blue. There is no blue pigment; we see the light most efficiently scattered and reflected by the distribution and nature of the melanin in the eye. So the color of the bluest eye is in the eye of the beholder, not the eye of the holder.

Genomic studies (DNA sequencing) reveal that genes controlling the nature, synthesis, and distribution of melanin have undergone many mutations generating several forms of these genes. Some versions are associated with dark skins, others with lighter hues, some determining one eye color, others a different shade. Similar considerations apply to hair color, for studies have found that that many genes collaborate to determine hair color. Not surprisingly, some of these are the same as those that influence skin and eye color. It is now possible to look at the DNA sequences of a thoughtfully selected set of genes and draw surprisingly reliable conclusions about the skin, hair, and eye color of the bearer of these genes. Recall from an earlier chapter that based on studies of their DNA sequences, it was possible to infer with some degree of certainty that the long extinct Neanderthals that preceded us as the dominant variety of human had light skin and light eye shades. Some were probably redheads.

Humans arose in Africa, and we have spent three-quarters of our time on Earth in its mostly sunny environments. Challenges accompany life under strong sun. All other things being equal, in equatorial regions it is hot under the sun. Becoming "naked apes" was a heat-coping mechanism evolved by humans. Parting company with our great ape ancestors, we shed most of our body hair and increased our number of sweat glands, gaining the advantages of evaporative cooling that accompanies sweating. As we will see, this evolutionary adaptation satisfied requirements for vitamin D but compromised levels of folate (vitamin B9). Each of these agents plays an indispensable role in promoting and maintaining health. Vitamin D is essential for many functions including the absorption of calcium from the intestine and a deficiency in this vital mineral results in an inability to build and maintain strong bones. Vitamin D deficiency in early life is responsible for the developmental disease, rickets, characterized by the strikingly bowed legs developed by children who do not receive amounts of this vitamin necessary for the uptake of adequate amounts of calcium. Folate, also known as folic acid, is essential for the synthesis of the building blocks that make up DNA and RNA, the body's key molecules for the storage and expression of genetic information. Folic acid is widely available in nature and diets that include leafy vegetables, meats, eggs, and even beer provide good sources of this essential nutrient.

In contrast, dietary sources of vitamin D are few and limited to oily fish and the flesh, especially the livers, of large mammals. Humans and many other animals satisfy the body's vitamin D requirement by a sort of photosynthesis that produces it in the skin by a process that is absolutely dependent on sunlight. However, folate is sensitive to light and in bright sunlight can be degraded in the skin. Equatorial Africa's bright sun provided ample sunlight for the support of vitamin D synthesis but would have put naked apes at risk of folate deficiency—unless they evolved darker skins. The first members of the line of apes we call humans probably had the same white complexions as chimpanzees, our last common ape ancestor. As Nina Jablonsky and others have explained, evolution had to find the range of skin color that admitted enough light to power the synthesis of adequate amounts of vitamin D, but not enough to degrade folate. Evolution had to find a Goldilocks level—not too little, not too much, but just right. In equatorial Africa this required a dark skin.

When humans migrated out of Africa and into areas of Eurasia, where sunlight was fainter, the evolutionary shoe was on the other foot and there was an opportunity for selection to favor lighter skins. Under these conditions, lighter skin tones facilitate the penetration of sunlight to the vitamin D synthesizing layers of the skin to provide adequate levels of this vitamin without compromising levels of folate. In some individuals, the skin has so little pigment that exposure of just the face for a few hours each day allows absorption of enough sunlight to satisfy the vitamin D requirements of those individuals. A survey of human skin color, just before the European voyages of discovery resulted in new waves of migration, would have found a gradation of skin color from equatorial regions northward, with darker-skinned populations around the equator and lower latitudes and progressively lighter skins as one looked northward. Observation of this gradation of skin color from dark to light, directly related to solar intensity, provided a validation of the dependence of vitamin D synthesis on the capacity of sunlight to penetrate the skin.

However, Eskimos of North America and the Inuit of Greenland and the Eurasian Arctic would have presented a clear departure from the pattern observed elsewhere. Indigenous people of these areas have complexions that are significantly darker than would be expected in regions of very low average solar intensity. These darker skins would compromise the capacity of Arctic peoples to synthesize vitamin D. In this instance, the vital imperative of maintaining adequate levels of vitamin D is satisfied by cultural rather than biological adaptation. The culinary culture of Eskimos and the Inuit include the oils of ocean fish and the livers of sea mammals—seals, whales, otters—which are all rich sources of vitamin D, and their consumption satisfies the requirements for this vitamin, bypassing the need for light skin to thrive at these arctic latitudes. In the industrialized world of today, the supplementation of foods with vitamin D and the availability of vitamin D-containing vitamin pills is widespread and provides a technological, as opposed to constitutional, way of satisfying this nutritional requirement.

At some latitudes, however, such as those found on the Eurasian land mass, solar intensities vary greatly during the year, with long, bright, summer days and short, gray, winter ones rendering a skin tone appropriate for one season of the year not optimal for others. To protect against folate degradation during the times of year when solar intensities are higher, tanning, especially in lighter-skinned populations, provides a folate-sparing, somewhat darker skin tone during the time of year that has higher intensities of sunlight and allows a lightening of the skin during the months of lower sunlight. The biology of tanning is beautifully regulated to achieve this end. In the skin of individuals capable of tanning, once sunlight reaches or exceeds a threshold intensity, it induces the synthesis of melanin, darkening the skin, providing the sun shield necessary to avoid folate degradation. As levels of sunlight wane with the change of seasons and shortening of days, levels of melanin fall, the tan fades, and the individual's skin tone returns to a shade better suited to synthesis of vitamin D in folate-sparing lower levels of sunlight.

This tidy account of the evolution of skin color provides important and interesting insights but is misleading without additional geographical and historical context. Geographically, this simple view has light skins in the higher latitudes and darker ones in those within and near the tropics. Historically, it suggests populations left Africa with dark skins and evolved lighter ones as they migrated into Eurasia. As it turns out, a much richer and more complex picture provides a more accurate view of reality.

Recall from an earlier chapter that the uniformity of light skins immediately prior to the "voyages of discovery" in the part of Western Eurasia we call Europe was a relatively recent development. Six thousand years before Columbus, significant parts of this area and the British Isles were populated by dark-skinned hunter-gatherers. This population and its similarly dark-skinned ancestors had inhabited this relatively low solar intensity area for tens of thousands of years. How they got adequate amounts of vitamin D is a question. Perhaps the answer for them was diet, the same solution that continues to work so well for Eskimos and other indigenous populations of the Arctic. Early Europeans may have gotten their vitamin D from the flesh and innards, particularly the livers, of animals they caught, killed, or scavenged as carrion.

In addition to the geographical fly in the ointment presented by dark skins in Europe and darker-skinned Arctic populations, recent detailed studies, particularly the DNA sequencing of diverse African populations by Sarah Tishkoff and her associates investigating the genomics of skin color distribution in Africa, has revealed the genetic basis of the wide variation in skin pigmentation seen in Africa. There one finds some of Earth's darkest skins and, even leaving out the North African populations of Egypt and North Africa, skins among the San people of South Africa and some others populations, too, that are much lighter. This more complex picture encourages the view that skin color is only one of a number of evolutionary solutions our species has deployed to cope with the challenges presented by the problems of vitamin D synthesis and the preservation of adequate levels of folate synthesis.

LACTOSE TOLERANCE—NOT *EVERY* BODY NEEDS MILK

Lactose tolerance, the ability to digest and benefit from the food value of milk sugar beyond infancy and throughout life, is another trait that varies in frequency among racial groups. It has attracted considerable interest because it appeared relatively recently in human history and provides a clear demonstration that humans are still evolving. Variations in lactose tolerance among different populations provide a striking illustration of the interplay of biology and culture on the evolution of a particular trait. It was the development of animal domestication that set the stage for the emergence of lactose tolerance. Like the invention of agriculture, animal domestication was a cultural advance that had great impact on the populations that practiced it. Practitioners of animal husbandry didn't have to find, run down, and kill sources of animal protein; their food animals lived and traveled with them. Farmers and pastoralists quickly discovered another benefit of rearing mammals—they could be milked. Milk, a food so complete that it is the sole source of nutrition for newborn and young mammals, provides protein, fat, carbohydrate, and several vitamins and essential minerals.

Lactose, also called milk sugar, is the carbohydrate present in milk and contains abundant energy locked up in its structure. Unlocking the structure of this complex carbohydrate requires the action of the enzyme lactase to break it down into simpler sugars that can be absorbed from the intestine and used by the body. Although all mammals are born with lactase and are able to digest the lactose present in their mother's milk, early in life after leaving infancy, the ability to make the enzyme lactase wanes and the ability to digest lactose is lost. This is true of all mammals—cattle, cats, dogs, whales, whatever—except for some humans, who continue to make lactase into and throughout adult life. This trait is called lactase persistence and people who have it can consume milk without suffering the discomforts of lactose intolerance, the symptoms of intestinal discomfort that accompanies the inability to digest milk sugar.

Our solitary position as the only mammal that includes some populations that are adult consumers of milk is more than just lucky happenstance. Some years ago it became apparent that there were high frequencies of lactase persistence in populations that have a long cultural history of dairying, the practice of collecting and consuming, in some form, the milk of their cattle. All other things being equal, in dairying populations, those who retain the ability to digest lactose, the lactase persistors, gain an evolutionary advantage over those who can't, the nonpersistors. We now know that this trait is the result of a mutation that causes the production of lactase to continue into and throughout adult life instead of shutting down after infancy and early child-hood. Those who had the mutation could derive greater food value from milk consumption and as a consequence had a slight advantage over others who lacked the trait of lactase persistence. This gave them a small, but real, reproductive advantage. In the language of evolutionary biology, they were more fit and increased fitness led to populations with increased percentages of lactose-tolerant individuals. This is why populations with a history of dairying show the highest frequencies of lactase-persistent individuals. Al-though frequencies vary widely around the world, only about one-third of the overall human population is lactose tolerant. The other two-thirds are not. It is important to be aware that just determining whether or not a population is more or less lactose intolerant would not by itself identify that population as a racial group. Like skin color, lactose tolerance or lactose intolerance is not a racial trait. It is a trait that varies in frequency among racial populations.

SOME OTHER TRAITS
THAT VARY AMONG RACES

In addition to the traits used in our thought experiment, there are many others that have been identified as varying in frequency among the racial popula-tions listed on census forms of the United States. They include the color, form, and length of hair. Under the microscope, the hair of most people of East Asian ancestry is black and poker straight. Examination under the microscope reveals it to be oval in shape and have a regular pattern of medium-size black melanin granules that give it color. Those of European ancestry identifying as White have a variety of hair colors ranging from black to the very light blond of a few. Individual hairs are generally straight or wavy, oval or round, and have small to medium-size melanin granules that are responsible for whatever hair color is displayed. The majority of American Blacks have black hair fibers that are darkly colored by a high density of large melanin granules. The strong curl of individual fibers is responsible for the intertwined mat of hair that differs sharply from the straight or curly mats of Whites and Asians.

Also familiar are the differences in eye cast between East Asians and American Blacks and Whites. Many East Asians have an almond-shaped eye cast due to the presence of a fold in the skin of the upper eyelid (the epicanthal fold) where it approaches the inner corner of the eye. This structure is lacking in most White and Black Americans. Interestingly, though present in populations thought to be precursors of the Native American population, it is absent from Native Americans, suggesting its loss after the separation of these groups 15,000 or so years ago. Research has identified another mutation that arose in Asia that resulted in the formation of incisors with inner faces that are concave ("shoveled"). This modification is seen in many Native Americans. Since the mutation occurred around 35,000 years ago, it was already a genetic fixture of the population that migrated across the land bridge now submerged under the Bering Strait and founded the Native American population. The gene targeted by this mutation is one that affects a number of processes in addition to dentition. These include increases in the number of sweat glands and perhaps a reduction in breast size. As might be expected given our current tendency to fetishize breasts, this last possibility, albeit a speculative one, has attracted a great deal of interest outside evolutionary and developmental biology.

This list of racially associated traits, already extensive, could be extended, but there is no need to do so, the picture is clear. There are a large number of genetically determined traits that vary in frequency among populations conventionally designated as races. Since we know that the frequency of some DNA sequences differs among the major racial groups identified on the US census form, we should expect that at least some genetically determined constitutional traits would differ among racial groups, too. The thing to bear in mind is that a diverse set of DNA sequences is used to identify a racial population. Also notice that since these indicators are statistical and apply to populations, an individual member of a racial population may possess or lack any one or even several of these various characteristics. Race tells something about gene frequencies in populations but provides an unreliable profile of the genetic makeup of an individual member of a racial population. Of all the points made in this chapter, indeed in the book, this is perhaps the most important.

RACE AND CONCEPTIONS OF BEAUTY

As noted, people come in a staggering variety of skin tones, facial features, hair textures, hair colors, and body types, and some of these have strong associations with race. The notion of race was inspired by early biologists dividing human populations into groups they thought looked more like each other than like other groups. European classifiers, the inventors of the notion of race, proceeded to declare that Europeans (Whites) were superior in most regards, including appearance, to other humans. "Black is beautiful," an assertion of race pride first shouted in the 1960s, and a colorful reaction to Eurocentric standards of beauty, invites a closer look at race-associated traits and perceptions of beauty. While skin color remains the poster trait for invidious comparison, other traits, especially facial features—eye shape, noses, lips—and hair texture are influential determinants, too.

The whole issue of visible characteristics associated with race is complicated by several features that seem common to human beings in many cultures. One of these is a tendency to value whatever is rare or expensive, perhaps as an indication of resources that a bride might bring or a rich husband might afford. In areas where famine is common, there seems to be a preference for women who are at least chubby and often very fat, a preference that may work to their disadvantage as immigrants to countries and cultures with abundant resources, including cheap, high-calorie foods. Only in the affluent West is the near emaciation of some highly paid fashion models likely to be seen as an ideal of beauty rather than a call for compassionate help. Where the necessity of working in the fields betrays poverty, lighter skin is preferred, one of many acquired signals of affluence, like long-unbroken fingernails. This contrasts with wealthy Western circles where a tan speaks of moneyed leisure and advertises that, for some, color is a choice. Aesthetic preferences also influence childcare—for instance, how a child is carried or put down to sleep may influence posture and even head shape. Patterns and choices of adornment are uniquely human capabilities that are sometimes used to signify cultural preferences and status in ways that differ among racial groups. Thus, bodily decoration and tattooing are informative cultural signifiers that can tell us about diverse concepts of beauty.

A study to determine the world's most beautiful female and male faces conducted by England's University of Kent found that the top ten men and women all had one thing in common. They were white. The easy conclusion is that pale skin would be expected to be a central feature of beauty in a society so accustomed to "whiteness" as that of England. Not everyone is aware that lighter skin is also preferred in many nonwhite societies. These include those of China, Japan, Korea, and India, as well as much of the rest of east and south Asia. Unlike the paradoxical addiction to tanning seen in England, Northwest Europe, and the White population of the United States, it is common for Korean, Japanese, and Chinese women to go to some lengths to avoid tanning. These measures include avoidance of prolonged sun exposure and the use of sun-blocking lotions. India's Miss India contest, a beauty pageant that has run for more than sixty years, has yet to choose a dark-skinned winner. Ironically, Nina Davul Uri, a strikingly beautiful and talented young woman of Indian American heritage who was the winner of the 2014 Miss America contest, was the object of some negative commentary in

India where it was pointed out that her relatively dark skin tone would likely have been disqualifying for a Miss India title.

A preference for light skin tone extends across the Middle East into North Africa. Surprisingly, even "Black Africa" is affected by colorism. There is probably no better indicator of a preference for light skins than sales of creams and other preparations to lighten skin color. In India it might be Lakme Skin Whitening Cream in Korea, Etude House Spot Whitening Cream. In West Africa, an area including Ghana, Senegal, and Nigeria, skin whitening products, such as Binatone are a multibillion dollar industry. Aside from the cultural commentary delivered by the widespread use of whitening agents by nonwhite peoples, many users face damage to their skin and health.

Some of these products pose health risks because of the dangerous ingredients they contain. Mercury is an ingredient in some of these creams, hydroquinone or clobetasol propionate is an active ingredient in others. Mercury, a toxic heavy metal, can cause a variety of health effects including neurological damage. Hydroquinone, an agent that interferes with melanin synthesis, is also a carcinogen. Clobetasol proprionate, is a corticosteroid with side effects that include skin damage as a consequence of prolonged use. Recognition of the potential of each of these agents to cause adverse health effects raises concern about their use for the cosmetic and culturally motivated pursuit of lighter skin tones. However, the social forces driving the use of skin whiteners are powerful and pervasive factors encouraging the use of these products by much of the world's nonwhite population. Once societies associate lighter skin with desirable appearance, skin tone becomes a factor in many important areas of peoples' lives including mate selection, employment preferences, and self-image. This helps understand why Nigeria, Black Africa's most populous country, has the world's highest percentage of women using skin-lightening products, 77 percent according to the World Health Organization.

Even White populations don't escape colorism. If there is an aspirational look among Whites, it's the Nordic one featuring the iconic blond hair and blue eyes. While eye color is hard to modify cosmetically, hair color is readily changed. Chemistry offers up a color palette that includes a spectrum spanning a range from ebony to ivory and asks what you might like. A title from one of yesteryear's classic films claimed *Gentlemen Prefer Blondes*. Apparently customers do, too. Products that color or bleach hair to some version of blond enjoy brisk and durable sales. But biological blondes use genetics rather than chemistry to generate their pale locks. Although it is genetically determined, hair color is not determined by a single gene. Extensive surveys of DNA sequences from many individuals has revealed that dozens and dozens of genes influence hair color and more are suspected. At least eight regions of DNA affect the state and degree of blondness an individual displays. Even the casual observer of European populations and their

American derivatives notices that there are many more women with light hair than men. The easy explanation is that more women than men bleach their hair blond.

Color is not the only feature that members of some nonwhite populations artificially transform to make them more like those of White people. An almond-shaped eye rather than the wide one common in European populations is the eye cast of most members of the East Asian populations of China, Japan, and Korea. So it is telling that plastic surgery with the intent of changing eye cast is the most common plastic surgery requested in East Asia and the third most common (behind rhinoplasty and breast augmentation) requested by Asian American women. Despite the risks of infection, difficulty closing eyes, and injury to eye muscles, demand for this procedure, called blepharoplasty or "double eyelid surgery," remains high.

Julie Chen, a television personality and producer, underwent the procedure in response to comments from an agent and a TV news director warning that as long as her eyes had a certain "look" she would not advance to top billing. Realizing that what could have been viewed as offensive and racist comments were offered with constructive intent, she gave their advice serious consideration. Her decision was complicated by the divided opinion and misgivings within her Chinese American family. Some relatives were supportive. But others, understandably disturbed by the implication that there was something wrong with the shape of her eyes, opposed the procedure because they felt making such an alteration was a denial of identity. There is an obvious irony in viewing a feature that is anatomically normal for such a large fraction of the world's population as something requiring "correction." At any rate, undergoing blepharoplasty was not an easy decision for Julie Chen. Nevertheless, impelled by the pressures of her professional aspirations, she elected to have the procedure. It required two surgeries over the course of two years. She reported that each of these was quite painful. However, with the completion of the eye-altering cosmetic surgery, she experienced a significant upward acceleration in career mobility.

There are other ways in which one population attempts to artificially align some aspect of their appearance with those of another. The cosmetic transformation of broad or large (rhinoplasty) noses into narrow or smaller ones is practiced in some Asian cultures and of course has been immortalized in the stand-up routines of yesterday's Jewish comedians. A particularly creative one has the recipient of a successful "nose job" extolling the result as "a thing of beauty and a goy forever," inspired by the even more famous original "A thing of beauty is a joy forever" from Keats's poem, "Endymion." Though facetiously made, this last point is a reminder that vast sums are also spent in Western societies to alter physical characteristics like crooked teeth or oversized noses.

Finally, there are the determined assaults many Black Africans and their emigrant African American descendants inflict on their kinky hair with the goal of making those twisty coils poker straight. Older technology employed a metal comb brought to flesh-scorching heat, often in the gas flame of a kitchen stove, to straighten the hair. Many years ago the straightening comb was complemented or replaced by the introduction of a chemical treatment employing highly caustic alkali ("lye"), called "konking." Recent years have seen the lye replaced by formaldehyde. These chemical treatments do straighten hair, but the highly corrosive alkaline-based ones progressively damage hair and formaldehyde-based treatments depend on the action of a cancer-causing chemical. There was a period during the '60s and '70s when African Americans embraced biology and, for a time, widely adopted a style known as the "Afro" in which a halo of trimmed hair, retaining whatever kink an individual's genome dictated, was proudly worn. Perhaps the late, then young and vital, Michael Jackson displayed the most iconic of Afros. However, the mature years of this eerily gifted but tragically flawed genius saw him deploy every available code-switching technology, including those for modification of hair texture, in an effort to transform his phenotype from Black to White, a strange and misguided Pygmalion to himself.

WHO WILL BE THOUGHT BEAUTIFUL IN 2090?

It is not difficult to suggest motivations for the conscious and unconscious acceptance of White features as aspirational standards of beauty and appearance by so many in nonwhite populations. Increasing in strength since the fifteenth century and reaching a rising plateau of power by the nineteenth century, the dominance of Western culture (a collection of European White cultures) was a global reality, rich, powerful and beyond successful challenge until the end of the twentieth century. Willingly or grudgingly, the rest of the world looked to the West for standards of economic development, benchmarks of technological development, and advancement at the frontiers of scientific achievement. It was natural and expected that cultures of color, the rest of the world, would associate "Whiteness" and its hallmark traits with wealth, knowledge, and power. By extension, in the eyes of many nonwhite beholders, White faces were beautiful faces and set standards for beauty.

While 500 years of ascendance and eventual dominance may seem a long time, it is a brief interval in the long history of the modern humans, and we are seeing what promises to be an epochal change in relationship of Western and Eastern civilizations. The rise of China, perhaps signaled and prefigured by Japan's precocious twentieth century rise, is an important surrogate for the rise of other Asian powers. We have seen China again stand up in the late twentieth century and even accelerate its rate of development in the early twenty-first century to now approach a position of parity with the West. Many envision the possibility, even the likelihood, of this East Asian giant eclipsing the West as the century matures. In the light of recent developments and those expected in the near future, it will be surprising if "Whiteness" and the physical characteristics associated with it continue to set global standards of beauty.

Chapter Four

Race and Medicine

"Peculiar elongated and sickle-shaped red blood corpuscles in a case of severe anemia."
—*Archives of Internal Medicine,* November 1910 issue
Title of the paper by Dr. James B. Herrick that reported the discovery of sickle cell disease, the first race-associated inherited disease

In September 1904, Walter Clement Noel, an Afro-Caribbean, arrived in New York on the SS *Cearense.* He had been sent to the United States by his wealthy, landowning family in Granada to learn dentistry. From New York, he traveled to Chicago where he entered the Chicago College of Dental Surgery. The first several weeks went fine, but shortly after Thanksgiving, he began to feel increasingly less well, and after the Christmas break, feeling weak and suffering dizziness, he sought medical help at the nearby Presbyterian Hospital. Over the next few years, his periodic bouts of illness were treated by a medical team led by Dr. James Herrick who diagnosed him with anemia. During this time, they made the striking discovery that his blood contained, in Dr. Herrick's words, "peculiar elongated and sickle-shaped" red blood cells. Looking back, this was the moment when Dr. Herrick discovered sickle cell disease (SCD).

Subsequent research would find that SCD is caused by the inheritance of genes that encode an aberrant form of the oxygen-carrying protein, hemoglobin. Sickle hemoglobulin compromises the ability of red blood cells to carry out their indispensable function of transporting oxygen from the lungs to all of the body's organs and tissues. Those afflicted by SCD suffer a devastating syndrome of pain, anemia, increased likelihood of infections, and stroke. They have a decreased lifespan, and although it can be managed by various interventions and some of its effects temporarily relieved by treatment with the drugs hydroxyurea and L-glutamine, there is no widely available cure.

However, SCD is one of the best and most intensively studied genetic diseases. Some of the most advanced medical strategies, including gene therapy, are under consideration for the treatment of this devastating disease.

Sickle cell disease, once known as sickle cell anemia, is inherited and occurs in much higher frequency in some races than in others. Almost 80 percent of the world's nearly 4.5 million people afflicted with sickle cell disease are members of sub-Saharan Black populations. The other 20 percent or so are found in other populations, such as those of the Arabian Peninsula, India, and populations, such as African Americans, descended from sub-Saharan Black populations. Sickle cell disease, the first genetically determined disease definitively associated with race, underlines an important lesson we have covered earlier. Although the frequency is much lower than in Black populations, there are Italians, Greeks, and members of other Mediterranean populations, broadly lumped as White, afflicted with SCD. Because they do not have the genetic makeup required to produce the disease, most Blacks in the United States, about 997 out of 1,000 (99.7 percent), do not now have and will never have sickle cell disease. This is another example of a trait indisputably associated with race but far from uniform in its signature racial population. One lesson for those who practice medicine is that knowledge of the race of a particular individual does not allow one to conclude the presence or absence of a race-associated trait. Physicians will find that most of their Black patients will not have sickle cell disease and yet, though unlikely, there could be a White person sitting in their waiting room who is afflicted with this disease.

SOME DISEASES ARE MORE COMMON IN SOME RACIAL OR ETHNIC GROUPS THAN OTHERS

Although the first genetic disease determined to be more prevalent in some populations than others, SCD is just one of many inherited diseases whose frequency varies among populations. Cystic fibrosis is a disease that primarily affects individuals of European ancestry. In these populations, about 1 in 25 people are carriers, and the disease affects about 1 in 3,000 infants. Carrier frequencies in African and Asian populations are much lower. Cystic fibrosis primarily affects the lungs but may also impact other organs, such as the pancreas and the liver, and periodic difficulty with breathing and frequent lung infections are often seen in those afflicted with cystic fibrosis. Thalassemia, a disease that, similar to SCD, targets the vital oxygen-carrying protein of red blood cells, is more prevalent in such southern European populations as that of Greece. The gene determining Tay-Sachs disease, a devastating and fatal disease of early infancy, occurs at higher frequencies in Ashkenazi Jewish populations than in other Jewish and non-Jewish populations. All of these diseases have one factor in common. Each is caused by a defect in a single gene and so is dubbed a monogenic disease. Although there are thousands of diseases caused by defects in a single gene, most of them are quite rare, and many have an incidence of less than 1 in 100,000. Diseases that are caused or influenced by many genes, known as polygenic diseases, are by far the most important for human health and include some forms of the following: cancer, diabetes, high blood pressure, obesity, asthma, and various auto-immune diseases such as rheumatoid arthritis and lupus. Although these diseases are to varying degrees influenced by genetic makeup, they are also strongly influenced by the environment.

Culture, a broad and overarching term, includes how we regard, exploit, and impact our environment, the tools and strategies we use as individuals and societies to make a living, and the countless features of what we might call our lifestyle. Therefore, culture can greatly influence the course and severity of diseases that involve the interaction of genetic and environmental factors. Pervasive and encompassing, culture emerges as a major determinant of wellness and illness. Even though genes do sometimes play significant roles in major diseases, it is not surprising that cultural differences in behavior and other environmental factors, rather than genetics, are usually the most important generators of differences in health status and disease outcomes experienced by different racial groups.

While acknowledging the central role of culture, the interaction of genes and the environment can be an important and, sometimes, a key factor in the selection for certain genes in one population and selection against them in another. The gene for the sickle cell trait mentioned earlier illustrates this point well. We now understand that an environmental factor, the scourge of malaria, has been a powerful agent for the selection of genes that lessen the severity of its effect on an impacted population. Malaria is a severely debilitating disease that even today causes the death of over 400,000 annually in the world's malaria-affected areas.

Most of us are aware that malaria is caused by a parasite that is spread by some species of mosquitoes. Malarial parasites introduced into the bloodstream by mosquito bites gain entry into red blood cells where they multiply. However, their growth and multiplication are optimal in red blood cells containing the normal form of hemoglobulin. In people who have a copy of the normal gene for hemoglobulin and a copy of the gene for sickle cell hemoglobin, the red blood cell has a mixture of normal and mutant hemoglobulin. This compromises the growth and slows the multiplication of the parasite but does not harm the health of the human host who carries just a single copy of the sickling gene. In fact, should they be infected with malaria, those who carry the trait experience a milder form of the disease. Consequently, before the development of modern medicine, in areas where malaria is endemic, those who were carriers of the sickle gene had a survival advantage over those who were not. In such an environment, despite the tragedy of some individuals suffering sickle cell disease or other maladies, carriers of protective traits have a survival (and, therefore, reproductive) advantage. In spite of the costs of conceiving a child who will have SCD, the costs of malaria infection are even greater in an area where malaria is endemic, so there is a net benefit. In populations that have a long history of living in areas where there is a significant risk of malarial infection, we see a selection for genes that dictate the production of modified hemoglobulins that render red blood cells less hospitable to the growth of the malarial parasite. In populations of Africa impacted by malaria, sickle cell hemoglobulin is the major protective form of the protein seen. In addition to sickle cell hemoglobulin, in Africa and in other parts of the world, such as Southern Europe and India, impacted by malaria, one sees other protective hemoglobulin modifications and other genetically determined molecular defenses as well. However, as the following example shows, in environments where the risk of contracting malaria is low or nonexistent, the costs of protective genetic liabilities such as the sickling gene become very high since they confer no benefit.

Hitching a ride with the slave trade, the sickling gene traveled to the United States and the rest of the Americas. Today about 1 out of 13 African Americans carries a copy of the sickle trait gene. Since it takes two copies of the sickling gene to develop SCD, having only a single copy is a benign condition, and most are unaware of their carrier state, but here in the United States, malaria has not been a significant health problem for more than a hundred years. Furthermore, if for some reason there were an isolated US outbreak, a combination of technologies of insect control and public health measures would intervene to curb the outbreak and prevent a broad epidemic. Also, malaria can be treated with drugs of varying effectiveness. Therefore, the sickling gene that was at one time a helpful, if biologically costly, defense against malaria is now a liability, and when present in two copies, it is a life-threatening burden.

Since about 1 of every 13 African Americans carries the sickling gene, about 1 of every 169 marriages ($1/13 \times 1/13 = 1/169$) among members of this population will be between sickle cell trait carriers. Each of the babies born of such unions has 1 in 2 chance of inheriting the sickling gene from the father and a 1 in 2 chance of inheriting it from the mother; therefore, the net chance of inheriting a copy from both mother and father is 1 in 4 ($1/2 \times 1/2 = 1/4$). The frequency in the African American population of infants born with two copies of the gene encoding the sickle cell trait is 1 out of every 676 ($1/169 \times 1/4 = 1/676$) infants. Since these children harbor two copies of the sickle cell gene, they will suffer from SCD. Although there have been some helpful advances in the management and treatment of sickle cell disease, there is still no practical cure. It is hoped that recent developments in genetic engineering may provide a path to correction of the mutation responsible for the disease. However, such treatments will lie in the future and be expensive and difficult to apply widely. Fortunately, as we will see in what follows, a practical and inexpensive pathway for the dramatic reduction in the incidence of sickle cell disease already exists and could be taken now. The shape of that pathway comes from the experience of another ethnic group and emerges from the following story.

The Jewish people are an ethnic group that includes three major sub-groups: Sephardic Jews, Middle Eastern Jews, and Ashkenazi Jews. The majority of American Jews are Ashkenazi. Many years ago, Warren Tay and Bernard Sachs independently described a fatal disease, subsequently dubbed Tay-Sachs disease (TSD), that appeared in infants and featured loss of motor function, blindness, and progressive degeneration of the brain. Although it occurs with the highest frequency in the Ashkenazi Jewish population, like most other genetic diseases, it occurs, albeit in lower frequencies, in other populations too. Research has traced its cause to the presence of two copies of a mutated form of the gene that encodes beta-hexosaminidase A, an enzyme vital for the proper processing and storage of a cell membrane lipid. Fortunately, a simple and inexpensive biochemical assay can identify carriers of the gene who have only a single copy of the defective form of the gene and suffer no ill effects.

About 1 in 30 Ashkenazi Jews is a carrier of the TSD gene. As we saw in the discussion of SCD, the laws of genetics tell us that 1 in 4 of the offspring of mating between carriers of the TSD gene will receive a copy of the defective gene from the father and a copy from the mother, giving the baby two copies of the TSD gene, a fatal dose that will doom it to progressive brain degeneration and, in the majority of cases, death before the age of three. If the frequency of carriers in the population is 1 in 30, all other things being equal, 1 in 900 matings ($1/30 \times 1/30 = 1/900$) will be between TSD carriers. Since a quarter of the offspring conceived from such matings receives two copies of the TSD gene, about 1 of every 3,600 ($1/4 \times 1/900 = 1/3,600$) children will have a genetic constitution that results in Tay-Sachs disease.

With the passage of time and continued research, more inherited diseases seen at greater frequency in Ashkenazi communities than others were discovered. All of these diseases are rare. Their approximate frequencies range from approximately 1 in 900 to 1 in 60,000. Using sensitive and readily available biochemical and genomic tests, screening methods have been developed that allow otherwise healthy carriers of these diseases to be identified. Although the risk of an individual of Ashkenazi descent being a carrier for any one of the eight or so diseases of greatest concern is small, the likelihood of a member of the Ashkenazi ethnic group being a carrier of at least one of these eight or so diseases is uncomfortably high, about 1 in 4 to 1 in 5. Consequently, broadening the screening in Ashkenazi populations to include all of the eight or so of the diseases mentioned is now recommended.

A PROVEN APPROACH TO LOWERING THE INCIDENCE OF SOME INHERITED DISEASES

During the 1970s, a door-to-door campaign to educate people on the benefits of screening for Tay-Sachs began in some Jewish communities. Informal discussions and educational presentations to community organizations and religious groups outlined the benefits and removed the stigma that had been associated with carrier status. The initial success of this grassroots effort inspired the creation, in the Ultra-Orthodox community of Dor Yeshorim, called The Committee for Prevention of Genetic Diseases. Dor Yeshorim arranges anonymous testing of young adults. Samples taken from each individual are tested to determine carrier status for Tay-Sachs and, more recently, several other genetic diseases known to be present in the Ashkenazi at problematically high frequencies. Based on the results, each donor is assigned a special number that is archived by Dor Yeshorim. When members of the community consider marriage, they get in touch with Dor Yeshorim, submit their respective numbers, and ask if they are at risk for conceiving offspring who would suffer from any of these genetically determined diseases. The expectation is that those prospective unions that pose a high risk will be abandoned.

This strategy has been a highly effective one, greatly reducing the incidence of these diseases among the Ultra-Orthodox, its most conscientious practitioners. In subsequent years, additional organizations have offered testing providing a wider and more liberal choice of pathways for reducing the incidence of these devastating diseases. The success of these programs is strikingly illustrated by the fact that Tay-Sachs disease, once seen at highest frequency in Ashkenazi populations, has been impressively reduced. In fact, these approaches have been so successful that today, here in the United States, most of the children afflicted with Tay-Sachs disease are born to couples that are not Jewish!

Earlier, we looked at sickle cell disease, a serious and life-shortening genetic disease that results when the offspring of healthy carriers are dealt two sickle cell genes by the genetic lottery. Affecting almost 100,000 mostly Black Americans, this debilitating disease does not yet have a practical cure. Perhaps the success of the Ashkenazi community in lowering the incidence of Jewish genetic diseases suggests an approach that could be adopted by African American communities to lower the frequency of sickle cell disease. Fortunately, there is an inexpensive and simple test that readily and quickly identifies carriers of the sickling trait. There are many religious, fraternal, and community organizations that could sponsor and encourage members of the Black community to undergo testing. There are also many figures in the world of entertainment and sports that command the attention of large segments of the Black community. Recruitment of some of these figures for an educational campaign could be especially helpful in reaching teens and young adults since their voices and especially their Instagram and Twitter accounts might be especially persuasive.

If sickle cell trait screening were broadly adopted by Black communities across the nation, following testing, couples that learn they are both SCD carriers and are healthy but with the potential to conceive children with a genetically determined disease should be informed that they have a variety of options. They can decide to enjoy a life of love and companionship with each other but elect not to have children. For others, adoption provides a way to avoid imposing the 25 percent risk of SCD on biological offspring but still experience child-rearing and provide a supportive home for the children they love and parent by choice.

If prospective parents have the financial means to do so, artificial insemination of the mother with sperm from a donor that is not a carrier could be used. Alternatively, in vitro fertilization allows eggs from a donor who is not a carrier to be fertilized in vitro with sperm from the father. Either of these approaches gives the child a 50 percent chance of being a carrier but assures no SCD. Another higher-tech alternative is a combination of in vitro fertilization and the use of technologies that allow the embryos to be genetically tested prior to implantation in the mother, selecting only embryos for implantation that tests prove are not at risk for SCD. A riskier course is to conceive and then rely on some form of prenatal diagnosis, which can be done as early as ten weeks into the pregnancy, to identify offspring that bear two sickling genes. In the face of a determination that the child is destined to be born with SCD, some parents will choose to end the pregnancy. Those that elect not to terminate should expect their child to be afflicted with SCD and prepare for this likely outcome by having plans in place to provide their child with the most current supportive and management therapies available.

SOME CANCERS APPEAR AT HIGHER RATES IN SOME RACIAL POPULATIONS THAN OTHERS

Cancer provides some vivid, if tragic, examples of the roles played by ancestry, culture, and environment in the occurrence of some of its forms. *The Biology of Cancer*, Robert Weinberg's definitive text on cancer, points out that the incidence of different types of cancer varies widely among different cultures and in different parts of the world. As examples, consider that the world's highest incidence of stomach cancer is seen in Japan, the highest incidence of prostate cancer is in the African American population of the United States, and the highest rate of melanoma is seen in White Australians. Weinberg also notes that the world's lowest death rate from breast cancer is seen in China. Here in the United States, cancer killed 600,920 Americans in 2017 and will soon surpass heart disease as the major cause of death in the country. Of all ethnic groups in the United States, African Americans have the highest overall rates of cancer incidence and death. However, as expected from what was said previously, there are differences when one looks at incidence rates of some particular cancers in particular populations. Hispanics have lower rates of common cancers—lung, prostate, and colorectal—than Whites. If we look at the statistics for some cancers caused by infectious agents such as gastric cancers associated with infection by *H. pylori* bacteria, liver cancer caused by hepatitis viruses, and cervical cancer, almost all of which is induced by human papillomavirus (HPV), we find the highest incidence of these cancers in the Black population. Asians have the lowest incidence of most cancers but higher rates of stomach cancer. Many interacting factors contribute to disparities in cancer rates and other diseases as well. These are complex and include both behavioral and biological factors.

With respect to biological factors, although we know that genes affecting the initiation and course of some different types of cancer occur at different frequencies in different human populations, the major factors responsible for most of the differences in the incidence of various cancers are cultural rather than genetic. For example, *The Biology of Cancer* cites Julian Peto's account of a striking study of the incidence of stomach cancer in migrant and home island Japanese populations. In Japan, rates of stomach cancer are more than four times those of Japanese who have been in Hawaii for a generation or more. Interestingly, in the Japanese Hawaiian population, the rate at which this cancer occurs is almost identical to that of the White Hawaiian population. Since Hawaiian Japanese are genetically very similar to Japanese in Japan, what is different is their culture (i.e., how they live in their adopted homeland).

In Australia, we see genetics and culture intersect to afflict one racial group with a heavier burden of particular cancer than other groups. As already noted, the world's highest rate of melanoma, a life-threatening skin cancer, is found in White Australian populations. However, other darker-skinned Australian populations, such as the Aborigines native to the island continent as well as the growing population of immigrants from the Indian subcontinent, do not suffer elevated melanoma incidence. Genes endow Australian Whites with skin that has little melanin and, therefore, admits more of the sun's cancer-causing ultraviolet rays than darker skin, a feature that places the DNA of their skin cells at greater risk of suffering cancer-initiating mutations. The combination of genetically determined UV transparent skins and many activities, such as sunbathing, that chronically expose unprotected skin to the bright Australian sun produce a conspiracy of biology and culture that explains the higher incidence of this cancer among White Australians. Fortunately, an expensive and innovative medical technology is not necessary to significantly reduce the incidence of this cancer. If widely and conscientiously adopted, inexpensive behavioral modifications—use of protective suntan lotions and minimizing direct exposure of the skin to bright sunlight—would yield welcome and impressive declines in the number of people afflicted with melanoma, an avoidable race-related cancer disparity.

Regrettably, health disparities are a fact of life. In Norway, men have a life expectancy of eighty years. For men in Russia, the figure falls to seventy-one. Why is there such a disparity in the lifespan of these two populations? Since both populations are White or largely so, race is probably not an important factor, and we need to look elsewhere. A colleague of one of us who directs the lymphoma center at the Massachusetts General Hospital would say that the greatest cause of health disparities is economic. Thanks to a grant from the Paul Allen Foundation, each year, he and some of his colleagues donate a few weeks of their time to medical service in Botswana, a small and well-governed country in Southern Africa that lies, landlocked, just north of South Africa. Although Botswana's new medical school graduated its first class in 2017, the country will continue for years to suffer a profound shortage of doctors.

Given that there are too few general practitioners and pediatricians, it is not surprising that oncologists like this colleague and others similarly trained are especially few in number. In addition to a shortage of oncologists, cancer therapy in Botswana is handicapped by a shortage of medications, and there are not enough of the standard agents of chemotherapy and the genetically engineered agents essential for many of the newest approaches to lymphoma, such as immunotherapy. Costs put these agents and approaches out of the question. There is only one machine in the country for delivering radiation therapy, which, along with surgery and chemotherapy, is one of the mainstays of cancer treatment. Even though this generous band of volunteer physicians does what they can and despite the helpful generosity of the Allen Foundation, the disparities in cancer care and health care in general between Botswana and countries like the United States, France, or Japan are enormous. These disparities are rooted in economics and will remain facts of life until the Botswanas of the world have the financial resources required to build and maintain health-care systems adequate to meet their needs. However, there is no need to travel as far as Botswana to find impressive health-care disparities that could be significantly reduced with a greater investment of resources.

HEALTH DISPARITIES ARE
WIDESPREAD AND VARIOUS

Here in the United States, profound and conspicuous underinvestment in health care for American Indians underlies a wide health-care disparity between Native People and the White population. A glance back through history shows there is nothing new about such resource disparities. An 1890 annual report from the Office of the Commissioner of Indian Affairs states that doctors working with Indian populations were paid an average yearly salary of $1,028 while army doctors were paid $2,823, about 270 percent more. Using figures from 2014, federal expenditures for health care were just over $8,000 per capita in the general population of the United States, but only a little over $3,100 per capita was invested in the health and wellness of tribal communities. It is sobering to acknowledge that a great deal more is spent on health care for prisoners than for those living in tribal communities.

Again, using data from 2014, we find that the national average for health-care expenses for those incarcerated averaged about $6,500 per inmate, with California funding at the highest levels of support—$19,796 per inmate. The strikingly more modest $3,100 per capita average in tribal communities makes an uncomfortable indictment of our priorities and reveals an inability to adequately fund a wide variety of essential services for Native Peoples. Let's be clear, this is not an argument for reducing the quality of care in prisons; it makes a compelling case for making up the health-care deficits borne by tribal communities. The deficit includes hospitalization, specialized outpatient procedures, diagnostic imaging, laboratory procedures, and the emergency treatment and care of trauma patients. Tribal communities also recognize that an inability to fund mental health programs at adequate levels is a major contributor to problems of depression, suicide, and substance abuse that impact life in these communities. Suicide rates in tribal communities are the highest in the United States, with a rate 2.5 times the national average for those in the fifteen to thirty-five age group, which is an age bracket that includes adolescents and young adults, the human capital that is the future of these communities and our nation.

The gaps in the smiles of too many in tribal communities reflect a failure to fund access to dental services, despite our knowledge that poor oral health has deeper systemic effects far more important than the merely cosmetic. Data from the federal Indian Health Service shows that in tribal communities, over 80 percent of children between the ages of six and nine suffer from cavities while less than 50 percent of comparably aged children in the wider United States population have had cavities. Funding that provides more dentists and dental clinics on reservations is an obvious correction to the state of poor dental health in tribal communities.

Tight and inadequate budgets make it difficult to build and equip the health-care infrastructure needed in tribal communities. They are often in remote locations situated far from facility-rich, major urban centers. Consequently, these communities are especially in need of local medical facilities that can be reached without the requirement for those living on reservations to have the means and the ability to travel many miles to distant and unfamiliar places to secure treatment for themselves or their children. Recruitment of doctors and nurses to tribal communities has been especially challenging because they are handicapped by an inability to offer generous salaries and benefits packages. This situation is exacerbated by the growing and increasingly serious shortfall in the nation's production of primary care physicians and nurses. Furthermore, many of these sought-after professionals prefer to settle in major metropolitan areas, and it is difficult to persuade them to take their much-needed skills to small cities, towns, and especially reservations. It is clear that until budgetary problems are addressed, American Indians will continue to experience higher death rates than the nation as a whole. This mortality rate for Native Americans is 133 percent higher than the rate for the nation as a whole. It is especially high for maladies such as chronic liver disease (460 percent higher), diabetes (320 percent higher), and kidney disease (150 percent higher), and rates for some other ailments are also elevated. While acknowledging that economics is not the only driver of these disparities, it is a primary one, and more money would cure a lot of the ills currently afflicting the system.

At the age of five, one of us watched his father die. Susceptible to asthma attacks, this Black man in early middle age was seized by one of his periodic asthma attacks while in the midst of building a rock garden in the backyard of his family house in Kansas City. Over the course of the next couple of hours, his condition worsened, and he died. It was thought that a doctor's advice and assistance might have altered the course of events. This was at a time when you picked up a telephone and "called" a doctor, expecting your personal physician, little black doctor's bag in hand, to make a house call. Attempts to summon his physician and, after that call failing, attempts to reach other doctors known to the family were unsuccessful. All of them were in some kind of medical conclave and impossible to reach on that long-ago Saturday afternoon.

Although his wife made increasingly frantic efforts to reach doctors who lived much farther away, there was a doctor who lived in a big red brick house, close by, on the next block. However, it would not have occurred to her to run to the White doctor's house and get medical help. Not because she couldn't pay the bill, but because at that time, in the Kansas City of 1940, medical practice, like religious worship, was racially separated. Back then, racial lines were absolute. Black people were treated by Black doctors, went to Black hospitals, and, when they died from asthma, heart attack, old age, or whatever, they were buried in Black cemeteries. Prior to burial, their bodies were embalmed by Black morticians such as Watkins and Sons, a venerable institution that even today continues to serve Kansas City's Black community. Although unable to appreciate it at the age of five, watching this death, perhaps preventable, was the first encounter for one of us with a race-related health care disparity.

Three-quarters of a century later, it is still easy to provide dramatic illustrations of racial disparities in health care. The distinguished physician and thinker on health-care policy Michael Marmot has pointed out that if you get on a subway in the Southeast district of Washington, DC, and travel out into the suburbs of Montgomery County, for each mile you travel, life expectancy rises about a year and a half over that of the poor Black residents where you had boarded. Toward the end of your ride, you reach wealthy, largely White populations with life expectancies twenty years higher. Even though the de jure barriers of racially segregated health care and hospitals mentioned previously have fallen, de facto barriers of access, such as differences in the kind and quality of care available and delivered and in affordability, still conspire to produce striking inequalities. These play out as health disparities in Black, Hispanic, and, as explored previously, Native American communities across the United States.

Early in the twenty-first century, *Unequal Treatment: Confronting Racial and Ethnic Disparities in Healthcare* was published by the Institute of Medicine, an arm of the National Academy of Medicine. This landmark study, still a benchmark in the field, presented the results of an in-depth and searching study of health-care disparities. The study found that racial disparities in health care take many forms and demonstrated that they are a systemic and durable feature of our social landscape. The findings were notable.

The study reported variation by race and ethnic group in the rates at which some medical procedures were used. Disparities were found for procedures such as bypass surgery, dialysis, kidney transplantation, and the prescription of some cardiac medications; Blacks and non-White Hispanics got less. The study also found disparities in the administration of less desirable procedures, such as amputating the lower limbs of diabetic patients; Blacks and non-White Hispanics got more. Certainly, there was a time when there were many instances when a lower quality of care was knowingly, even deliberately, given to groups such as Blacks and non-White Hispanics. Although some vestiges of such overt discrimination may remain, there is good reason to conclude that today factors other than bigotry underlie much of the disparity that persists in our health-care system. To the extent that they arise from a lack of awareness among providers, administrators, and medical educators, one of the most productive approaches to their elimination is an insistence that evidence-based guidelines for treatment and care recommendations be universally and rigorously employed no matter the patient's race, ethnicity, or income.

Democratizing medical care will not be accomplished cheaply or easily. More of everything, especially money and outreach, will be required. More money will be needed because effective medications are often expensive, the infrastructure of buildings and equipment required for the practice and delivery of modern medicine is quite costly, and the costs of training and compensating highly competent medical personnel are high and will surely continue to increase. We as a nation must face what is obvious: providing high-quality health care to all of those who have not had it will impressively boost the nation's health care bill. We won't be able to escape the need for more outreach, which, carefully thought out and tailored to reach different constituencies, will be required to raise awareness and to inform new consumers about how to find and make the best use of whatever newly available opportunities for health care are created.

RACE AND PERCEPTIONS OF MEDICATION AND DRUG ABUSE

Race has also had a striking impact on whether some conditions are regarded as illnesses or manifestations of criminality. Opium and its close relative, heroin, are opioid drugs that have been abused for a very long time. Our response to opioid abuse has been socially refracted through prisms of race and class with surprising inconsistency. During the nineteenth century, White populations based in Europe and the United States sold opium to the Chinese. As use in China increased and the pool of buyers expanded, opium trafficking grew to be a highly profitable commercial enterprise. Unable to ignore the debilitating effects of opium addiction on growing segments of its population, China ordered a halt to the opium trade. Major powers in the West were unwilling to abandon this highly profitable drug dealing and ignored China's directive to put an end to their opium sales. When the Chinese deployed force to compel compliance with their edict, the response was the Opium Wars waged by England, with the enthusiastic backing of the United States. The superior military power of England was able to force China to acquiesce in the resumption of this highly profitable but socially corrosive and destructive trade.

Many will be surprised to find that some highly prominent American families, including icons of patrician America such as the Roosevelts, profited from this lucrative drug trade. The "Delano" in Franklin Delano Roosevelt was given in recognition of his grandfather, Warren Delano, who made a part of the family's foundational fortune in China by trading opium. The gardens of social progressivism that we associate with President Roosevelt were fertilized by the rich and malodorant manure of drug dealing. Today's drug dealers, plenty of whom are Blacks or non-White Hispanics, serve jail time and usually make far less than some nineteenth-century American patricians gained from their socially destructive but very profitable drug trafficking.

Heroin is a strong contender for the title of poster drug of the opioid epidemic. For generations, its sale and use in inner cities have been recognized as a persistent and socially malignant cancer afflicting many Black neighborhoods. In this context, the culture of heroin dealing and heroin use were regarded as manifestations of community dysfunction and social degeneracy. Transactions involving the sale or consumption of "homeboy" heroin were criminalized, and law enforcement was assigned to discourage the sale and use of this narcotic and others. The management strategies employed by law enforcement were routinely brutal, and the punishment administered by the criminal justice system for the sale or use of these drugs was harsh, characterized by frequent incarceration and long sentences.

The latest versions of opioid abuse, beginning with pharmaceutical versions of opioid pain medications and maturing to embrace the highs of heroin and the lethal dangers of fentanyl, appeared in communities far from the inner cities, communities that are White and often rural and poor. A major theme in the approach to the "hillbilly" heroin epidemic has been to place as much emphasis on medicalizing it as on criminalizing it. We finally see our society relying on the counselor as well as on the cop. Offenders are increasingly and appropriately regarded as clients rather than criminals. Regardless of the disparity in our approach to opioids when they were regarded as a social malady typical of inner cities rather than a community crisis of the larger society, the recent shift to an aggressive deployment of medical as well as law enforcement approaches is an important and welcome step forward.

SOMETIMES WHAT A DRUG DOES TO YOU AND WHAT YOU DO TO A DRUG CAN VARY WITH RACE

For a very long time, drugs, in some form or another, have been used for the treatment of disease. More than 3,000 years ago, medications were prescribed and taken for the treatment of illnesses in ancient Egypt. *The Ebers Papyrus*, a handbook of Egyptian medicine that dates back to 1550 BCE, lists hundreds of different prescriptions for the treatment of ailments. Some of these are for managing diseases as serious as diabetes, others for treating debilitating ailments such as severe gastric disturbance, and some were given as palliatives for the relief of discomforts as banal as toenail pain. In the many years since the medical lore of *The Ebers Papyrus* was penned, a great deal has been learned.

We now appreciate that the effect of any drug is determined by two things: pharmacodynamics, which is what the drug does to you, and pharmacokinetics, which is what you do to the drug. Effective drugs always modify some aspect of the body's physiology—ACE inhibitors lower blood pressure, chemotherapy inhibits cancer growth, opioids decrease the perception of pain, and Ambien, we hope more soporific than what you are reading, increases the likelihood that you will soon fall asleep. These are all pharmacodynamic effects, descriptions of how a drug modifies some aspect of the body's function. On the other hand, once a drug is administered, the body transports it, often modifies it, making chemical alterations in its molecular structure, and usually excretes what remains of the drug and whatever byproducts have been generated from it. In some cases, the body's modification of a drug may be slight or inconsequential; in others, it can be quite extensive. The body's clearance of the drug and its products might be slow, allowing it to remain in the tissues for extended periods; other times, drug clearance might be fast, allowing it to disappear from the body rapidly. The body's modification of a drug may significantly modify its effect on the body, in some cases making it inactive or in others converting it to a form that actually produces the therapeutic effect of the drug.

Many factors, often genetically encoded features of physiology, determine what the body does to a drug and how the drug will affect the body. Therefore, it comes as no surprise that research has identified many genes that influence pharmacokinetics and pharmacodynamics. Some of these genes occur in different forms, each having slightly different DNA sequences from the others and therefore directing the formation of slightly different products. Such genes are said to be polymorphic. An example of genetic polymorphism was encountered earlier when different forms of a gene that determined whether or not one bears the sickle cell trait were discussed. Another familiar example of polymorphism is seen in the gene that determines ABO blood types. Each of the traits A, B, and O is determined to be a slightly different variation of the gene determining ABO blood types. Earlier, we discussed eye color. Some of the genes that determine eye color are polymorphic, and DNA sequencing has shown that the frequencies of these genes vary among populations.

Similarly, research and clinical experience have shown that many of the polymorphic genes determining how the body processes a particular drug or what that drug does to the body differ among populations. There are drugs that benefit or fail to benefit greater or lesser numbers of some populations than others. Polymorphisms in the genes that affect the uptake, distribution, metabolic processing, degradation, and, ultimately, the rate at which the drug is excreted can all render a drug highly effective in one population, so-so in another group, and ineffectual in yet another population. There are some drugs that show greater instances of toxicity in some populations than others.

These considerations can make it advisable and sometimes necessary to tailor the dosage and dose frequency of a drug to match the genetics of different populations better. However, populations are not uniform, and there can be a great deal of variation in drug responses within any particular population. As a result, within a racial population that on average responds this way or that way to a particular drug, at the individual level, we will find a spectrum of individual responses, often quite broad, that depart from those typical of the population. One size does not, necessarily, fit all, and physicians are reminded to regard their patients as individuals rather than thinking of their responses as absolutely determined by their population.

Race, after all, focuses on average differences of traits between populations. However, physicians treat individuals, not statistical averages. Although there are many instances in which race has proven a useful proxy for genetic differences that make it advisable to adjust the types and amounts of some drugs given to different populations, it is an inexact one. Just as many Blacks have brown or black eyes, and cystic fibrosis occurs at higher frequency in White than in Black populations, yet there are a few Blacks with blue eyes and some Whites with dark eyes, and Black patients afflicted with cystic fibrosis are not unknown. Ideally, medical decisions should be tailored to each individual with the goal being to get the right drug to the right patient in the right dose.

Pharmacogenomics is a rapidly developing field that eventually will make it possible in many instances to use an individual's DNA sequences to determine the likelihood that a particular drug will be efficient and safe. The day is approaching but has not yet arrived when the determination of medically relevant DNA sequences will be as routine as the determination of individual blood types. When that day comes, race will be abandoned as a proxy for genetic makeup, and we will know, with precision, many of the genetic predispositions, susceptibilities, and liabilities of each individual. For now, however, like age and gender, self-identified race remains a useful part of a patient's medical history. The following vignette illustrates the utility of knowledge of self-identified racial identity in the discovery and appropriate deployment of a medication for the treatment of heart failure.

A DRUG MAY WORK BETTER IN ONE RACE THAN IT DOES IN ANOTHER

More than 6 million Americans suffer from heart failure, a condition in which the heart progressively loses its capacity to pump enough blood to satisfy the body's requirements. This serious medical condition does not occur at the same rate in all populations of the United States. The rate of heart failure in Blacks is 190 percent greater than that of Whites, 130 percent greater than Hispanics, and 460 percent more than those of Chinese ancestry. Among those under the age of seventy-five, Blacks die of heart failure at rates that are almost twice as high as those seen in White populations. Because of its prevalence, there is great interest in the discovery and development of drugs for the treatment of this widespread and eventually life-threatening malady. Some years ago, a combination of two drugs, isosorbide dinitrate and hydralazine hydrochloride, dubbed BiDil by its manufacturer, was tested in a clinical trial to determine its efficacy for the treatment of heart failure. The results of the trial were disappointing. BiDil did not show efficacy for the treatment of heart failure in the general patient population.

However, subsequent reexamination of the trial data revealed a surprise. Although there was no benefit in the population that identified itself as White, some patients in the cohort that self-identified as Black did appear to enjoy a benefit from the treatment. In a subsequent trial that enrolled only Black patients, BiDil demonstrated that adding it to the conventional therapy for heart failure reduced the likelihood that one would die within the next year by 43 percent, a clear and impressive survival benefit. One result of this trial was Food and Drug Administration (FDA) approval of BiDil for the treatment of heart failure in Black patients. Another was the firestorm of criticism ignited by the FDA's decision.

Critics saw the decision as an undesirable, even dangerous, step toward race-based instead of individualized therapies. Those who view race as totally socially constructed and completely lacking in any biological reality whatsoever urged that therapeutic decisions must be completely divorced from race and made solely on the basis of physiology and genomics of individuals. Some of the objectors pointed out that no physiological or genetic basis for the differences in the drug's effectiveness in Blacks versus Whites had been established. Since even the most ardent advocates of race as having biological reality agree that few racial traits are borne by all members of a group, some patients categorized as Black would receive the drug but would not benefit from it because they lack responsive physiology. Just as importantly, a small number of patients who self-identify as White might just happen to possess physiological traits that would make them responsive to BiDil but would not receive prescriptions for the drug because they were White.

Some of those critical of the FDA approval of BiDil cautioned that the use of race in therapeutic decisions invites stereotyping, stigmatization, and labels of racial inferiority. A history that has many negative instances of deliberate delivery of inferior health care to minorities has not inspired these critics to embrace race-based drugs. However, now that the eye of the storm has passed, more than ten years of prescribing BiDil to treat heart failure in Black patients usefully has vindicated the decision of the FDA to grant this drug approval. It has saved lives and continues to help Black patients. In recognition of the contribution BiDil makes to the management of one of the

Black community's important health problems, the American Association of Black Cardiologists presented an award in 2016 to Arbor Pharmaceuticals, the manufacturer and distributor of BiDil.

Population differences in drug responses, genetically determined susceptibility to some diseases, and a variety of constitutional traits are well-established. Just as we expect to see physiological variation within a population, additional variation is seen between racial populations. Consequently, if we want to do the best we can to benefit the greatest number of patients and harm the fewest, testing of new medical procedures and drugs must not be limited to a single race. Just as clinical trials now take pains to include members of both sexes and people of different ages, it is increasingly accepted that they should enroll representatives of different racial and ethnic groups as well. Failure to do so will deprive pharmaceutical companies and medical practitioners of information they need in order to have the highest likelihood of increasing drug efficacy and safety in all of society's diverse populations. Furthermore, as the BiDil story demonstrated, such an approach allows discovery of drugs that may have little or no efficacy in a majority population but do provide significant therapeutic benefit to one or more of the minority subpopulations enrolled in the trial. While it has not always been the case, for some time now, guidelines issued by the FDA have strongly urged the voluntary inclusion of a broad diversity of participants in clinical trials.

Voluntary participation in medical research requires that the population from whom volunteers will be selected will trust the researchers. To build this trust, medical researchers are required to adhere to ethical standards that include the embrace of these four ethical principles: do no harm; do what is of greatest benefit to the patient; provide the information necessary for the patient to give informed consent; and treat all with similar needs in a similar manner.

MEDICAL RESEARCH HAS NOT ALWAYS ADHERED TO THE HIGHEST ETHICAL STANDARDS—THE TUSKEGEE EXPERIMENT

All of these ethical principles were violated in the Tuskegee Experiment, which stands as one of twentieth-century America's most egregious examples of a race-based violation of medical ethics. In the early 1930s, the US Public Health Service, a forerunner of the Centers for Disease Control and Prevention (CDC), wanted to demonstrate the need for syphilis control programs. It was decided that careful documentation of the course of this dangerous and devastating disease untreated would provide a powerful clinical demonstration of the awful and tragic consequences of allowing the disease to progress untreated. The investigators set their studies in Macon County, Alabama, because an influential consultant advised the following: "If one wished to study the natural history of syphilis in the Negro race uninfluenced by treatment, this county would be an ideal location for such a study."

Ultimately, 600 Black men, poor Macon County sharecroppers, two-thirds of whom had syphilis and one-third of whom did not, were enrolled in the study. These men were poor, had little education, and were not told that they were gathered as part of a study. Instead, they were led to believe that they had been selected to participate in a program intended to benefit their individual lives and health. They were told that free medical care, some meals, and free burial insurance would be provided to them by the federal government. Study subjects were never told that they had syphilis, a serious disease, dangerous to themselves and, because it is sexually transmitted, dangerous to their partners. Furthermore, because it can be transmitted from mother to child, it is dangerous to their offspring, too. Over the course of the next forty years, the subjects were observed and their medical histories meticulously updated with the results of periodic tests and observations made during visits to the clinic. Ingenious subterfuges were devised to hide the diagnostic purposes of some invasive and painful procedures. For example, subjects were told the spinal taps actually used to diagnose neurosyphilis were free treatments for some malady or other. During this time, illnesses progressed, some subjects died, forty wives were infected, and nineteen children were born infected with their father's diagnosed, but undisclosed, syphilis.

Among the many outrages of this study, which began in 1932, is the fact that by the late 1940s the powerful antibiotic penicillin was being used, often with spectacular success, to treat many previously intractable infections, including syphilis. However, it was never deployed during the study. In fact, the study leaders actively tried to prevent antibiotic treatment. In 1972, things came to an end when Peter Buxton, an investigator in the area of sexually transmitted diseases, found himself unable to persuade the CDC to stop the study by appealing through the usual bureaucratic channels. In angry frustration, he disclosed his story to the press. Sensational accounts appeared in the *Washington Star* and *New York Times*. These revelations quickly led to the study's termination and prompted congressional hearings.

Eventually, there was payment of ten million dollars to survivors and family members. President Bill Clinton issued an apology on behalf of the nation. Ironically, these experiments have been dubbed the "Tuskegee study of untreated syphilis in the Negro male" because the Public Health Service recruited the help and complicity of the then Tuskegee Institute, an influential and otherwise distinguished Black educational institution established by Booker T. Washington. The now Tuskegee University, to its credit, currently serves as a force for the reduction of health disparities. Rather than denying its participation in these experiments, Tuskegee acknowledges its then misguided role and is now home to the National Center for Bioethics in Research and Health Care at Tuskegee.

IN SUMMARY

An examination of race and medicine provides additional illustrations of the genetic and cultural diversity of humans. This diversity underlies some of the differences in the way different humans interact with the environment and the way the environment interacts with them. For the most part, we must look to culture rather than genetics to understand and correct the disparities in health care, some of them sketched here, experienced by underserved minorities in the United States and elsewhere. Only when the necessary cultural repairs are made will we see the extension of effective medical care to all.

Although culture plays a major role, an appreciation of the genetics of different human populations will also make an important contribution to the delivery of appropriate and effective medical care. Susceptibilities to malaria, the efficacy of some drugs, and the capacity to harvest the maximum nutritional benefits of some foods are influenced by versions of genes and groups of genes that differ among populations. Some specific examples have been reviewed in this chapter, making it clear that differences in gene frequencies, some of them associated with race, have clear implications for medicine and health. As we wait for the advent of inexpensive and widely available DNA-based assays for versions of genes that are associated with disease susceptibility and drug efficacy, self-identified racial identity will remain a valuable, if imperfect and sometimes troubling, surrogate for genomic tests.

Chapter Five

Race and Ability

"I advance it therefore as a suspicion only, that the blacks, whether originally a distinct race, or made distinct by time and circumstances, are inferior to the whites in the endowments both of body and mind."
—Thomas Jefferson, *Notes on the State of Virginia*

". . . all men are created equal . . ."
—Thomas Jefferson, *The Declaration of Independence*

Everybody knows that some people are just better at some things than other people. Sometimes the edge is more than training or diligence. You can't learn to create symphonies like Ludwig van Beethoven's, art like Pablo Picasso's, or plays like William Shakespeare's. You don't learn to have Einstein's profound insight into the nature of the universe. Even though there have been more than sixty years to study and to try, Boston's Red Sox fans know that nobody has learned to hit a baseball as well as Ted Williams. Gifts like those that blessed Beethoven, Picasso, Einstein, Shakespeare, and Williams are probably inherent and constitutionally based in DNA, in their genes.

Granting that some individuals are more gifted than others, are some races more able than others? Is it possible that some races have a higher frequency of individuals whose genetic makeup gives them an advantage in body or mind when compared to some other races? Thomas Jefferson thought so. Perhaps for a few things, the answer is yes, definitely, and with regard to those, Jefferson was right. The gene pools of some races do have higher frequencies of individuals who are likely to excel at some behaviors.

Basketball is high on the short list of such behaviors. Don't be too quick to dismiss excellence on the basketball court as trivial. Think about the social cachet that attaches to varsity members of high school basketball teams or to the respect accorded trash-talking stars of corner court pickup games. Moreover, if clear economic benefit is a preferred yardstick, Ivy League schools, state universities, and even many small colleges provide significant financial benefits to induce capable basketball players, both male and female, to wear the school colors and represent their institutions on the court. The average annual salary of men good enough to play in the NBA is an impressive 5.7 million dollars, far more than that of doctors, engineers, teachers, even MBAs. LeBron James and Michael Jordan would be comfortable comparing their net worth with those of members at many of the most exclusive golf clubs in the United States.

All other things being equal, as long as the basket is ten feet above the floor, taller players will have a significant advantage over shorter ones. David Epstein's fascinating book, *The Sports Gene*, points out that here in the United States, almost 1 of every 5 men between the ages of twenty and forty and who are seven or more feet tall are in the NBA! Height is strongly determined by genes, and the genes that determine height are not equally distributed across all racial groups. The gene pools of White and Black populations here in the United States produce a higher frequency of tall individuals than those of East Asian populations.

Not for the money but as an act of national will and a demonstration that almost anything can be made in China, a pair of unusually tall Chinese basketball players were encouraged to marry. The hope that they would produce exceptionally tall offspring was realized by the birth of their son, Yao Ming. He would grow to a height of seven-foot-six and be shaped by a rigorous training program from which he emerged as a highly accomplished basketball player who would eventually receive a ten-year contract worth over 90 million dollars to play for the Houston Rockets. However, even allowing for the occasional Yao Ming, here in the United States, more outstanding basketball players would be expected to emerge from its taller Black and White races than its shorter East Asian ones.

Consider track and field events such as the dash, where a gun fires, you tear off the starting blocks, and run as fast as you can for 100, 200, or 400 meters. Think, too, about the long jump, where you make a hard run at a marker and then leap for as many yards as you can. The great ones can fly through the air for distances that approach ten yards, almost a first down in one jump. The explosive bursts of speed required to win dashes or the sudden channeling of power to jump a long distance are facilitated by possession of particular physiological and anatomical endowments. Physiologists who have tried to understand why some people are better at these events than others have come to the conclusion that the critical traits appear to be found at higher frequencies in Black populations of West Africa and their descendants in the New World than in other populations. Notably, there is an overrepresentation of Blacks among the fastest sprinters and longest long jumpers. The casual excellence of Usain Bolt in the 100-meter dash and other events dependent on explosive speed makes him an exemplar of this point.

However, once we move beyond basketball and perhaps a few track and field events that rely on brief fast runs or long jumps, race-based prediction becomes unreliable. Consider the decathlon, a grueling ten-event contest that stretches over two days and includes four races of varying length, three different kinds of jumps, and three different kinds of throws. It provides a very broad assay of overall athletic excellence. Certainly, there have been several decathlon winners who are Black, but they are not the majority. Of the best-known winners, one is Jim Thorpe, an American Indian once dubbed "the world's greatest athlete." The majority of winners fall into the broad category we agree to recognize as White and include Caitlyn Jenner when she competed as Bruce Jenner. Turning our attention to soccer, arguably the world's most popular game, no race owns this sport of skill, speed, endurance, and maneuver. Moreover, there is baseball, America's most popular sport until it was muscled aside and blocked out of first place by pro football's Super Bowls, graceful passes, and bone-rattling, concussion-inducing tackles. Japan has shown a consistent ability to produce impressive baseball players at both the amateur and professional levels. Both Whites and Blacks, if they are big enough and strong enough, seem to have equal overall success at the brutal and brilliant game of football. Sports like soccer, American football, and baseball, do not place an excessive premium on any single attribute but instead require excellence in a wide spectrum of traits, many of which have strong genetic determination but which do not occur as a linked suite of genes concentrated in this race or that.

Today, it is difficult to find subscribers to President Jefferson's surmise of Black inferiority to Whites "in the endowments of body." Black athletic prowess is now widely acknowledged and sometimes fetishized. However, there is no need to go all the way back to President Jefferson to find a time when different assumptions held sway. In earlier times, when boxing—now a peripheral sport displaced by mixed martial arts (MMA), the brawl that masquerades as sport—was entering its glory years, there were followers of the "sweet science," as boxing was sometimes known, who were sure that Blacks, though powerful, could not match the strategic thinking of their White opponents. Also, there were some who wondered if Blacks possessed the reserves of courage and character necessary to sustain a fighter when the going got tough. Turning from boxing to baseball, many thought Blacks were unable to assemble the combination of skill, determination, and self-control necessary to succeed at major league baseball. Those whose perspective is confined to the late twentieth and early twenty-first centuries find it almost inconceivable that these concerns, long made moot by the success of Black athletes in the ring and on the diamond, were ever taken seriously. During the last half of the twentieth century, the outstanding performance of Blacks in these fields and other athletic pursuits increasingly became expected and routine.

However, note that this change of perception was not the result of closely reasoned and persuasive arguments, appeals for racial harmony, or the publication of scholarly essays arguing that there was every reason to assume that Blacks can compete at very high levels in all of these sports, matching and sometimes exceeding the accomplishments of White athletes. People know this is the case because the Black achievement in these fields became a part of their experience. Surely, President Jefferson's near-peerless intellect would have taken him to a similar conclusion if he had had the benefit of similar experiences.

WHAT ABOUT MENTAL ABILITY — A FAR MORE IMPORTANT ISSUE THAN ATHLETIC ABILITY?

Our species has no capacity for unaided flight. We cannot run very fast. We are slow swimmers. The most gifted pearl diver is limited to sea beds lying at modest depths and can only stay submerged for a few minutes. Still, there are no eagles that fly as high, no cheetahs as speedy, and no fish that skim the waters as fast or master the deeps as well as humans. The wings of birds, the quick twitch muscles of cheetahs, and the fins, gills, and streamlined body of fish endow each of these creatures with a narrow trademark competence. Each owes their particular competence to natural selection. So do humans. However, in this species, natural selection has assembled a highly compact computer—the human brain. It has a combination of power, capacity, and versatility that exceeds any computer now in production or on the drawing boards of computer engineers. It weighs about three pounds, and billions of human beings carry one around between their ears.

Because they were armed with these computers, many years ago, when a band of three or four hunters from a tribe of Plains Indians came up against a herd of bison, these huge, powerful, and fractious animals, some weighing almost a ton, were grossly overmatched. Buffalo meat was eaten and buffalo hides provided clothing and shelter for the tribe. In what we now call Mexico, Native Americans used these computers to guide the genetic transformation of stands of the wild grain teosinte that couldn't sustain a few deer into stalks of corn that nourish nations.

Humans are not the strongest, fastest, or fiercest animals, but they are, by far, the smartest. Intelligence is advantageous in all environments, and it is difficult to imagine an environment that would not select for it. Smart humans would always have an advantage because they would be more likely to learn from experience and make better guesses about what to expect. There are a variety of ways to be smart, and the distinguished psychologist Howard

Gardener famously recognizes distinct varieties, eight of which are outlined here:

1. Verbal-linguistic intelligence—includes the ability to understand, deploy, and manipulate language.
2. Logical and mathematical intelligence—the ability to identify logical or numerical patterns and to think conceptually and abstractly using mathematics or logic.
3. Spatial-visual intelligence—the ability to think in images, pictures, and shapes (including both real and imaginary entities) and to visualize these accurately and abstractly.
4. Musical intelligence—the ability to produce, perform, or appreciate rhythm, pitch, and timbre.
5. Interpersonal intelligence—the ability to determine and respond appropriately to the moods, motivations, or desires of others.
6. Intrapersonal intelligence—the ability to know oneself, including an awareness of inner feelings, beliefs, and thoughts.
7. Naturalist intelligence—the ability to recognize and categorize objects in nature such as plants, animals, and varieties of rocks and topological features.
8. Bodily-kinesthetic intelligence—the ability to control one's body movements and to handle objects skillfully.

Taken together, these capacities comprise all or most of the key package of mental abilities that made the rise of humans to a dominant position possible. Although not what many specialists in the field would accept as elements of a precise textbook definition of intelligence, these mental abilities combine to enable a variety of smart behaviors. Humans possessing most elements of this package will be aided in any environment. They will understand things, including themselves and those around them, better, more deeply, and more quickly. While it is easy to see evolution selecting for all of these abilities, perhaps even music, it is difficult to imagine an environment where one would not be better off and more fit with them than without them. Indeed, it is hard to see how being dumb increases fitness in any environment. Even a smart slave is more likely to understand how to accommodate to and make the most of an awful predicament than a dull one.

IQ IS A MEASURE OF A SUBSET OF IMPORTANT MENTAL ABILITIES

Some parts of this package lend themselves to quantitative estimation more readily than others. Psychologists have devised tools for assessing verbal-linguistic, mathematical, and logical competences as well as the ability to visualize and mentally manipulate real or imaginary shapes and images. These are called intelligence quotient tests or IQ tests. All of us have taken IQ tests and are familiar with their emphasis on verbal, mathematical, and logical mental abilities. Despite the generality implied by the name, IQ scores tell little about the mental abilities underlying social skills, self-awareness, kinesthetic abilities, or musical competencies. Certainly, the verbal gifts Bob Dylan (nee Robert Zimmerman) is blessed to possess would be detected by an IQ test. However, the extraordinary ability to creatively deploy sounds and rhythms, qualities that truly set him apart, would not. An IQ test would be blind to the field generalship, athletic poise, and passing elegance of the New England Patriot's legendary quarterback Tom Brady, and it would fail to register the otherworldly court sense and purposeful athleticism of LeBron James, but knowledgeable fans and struggling defensive players would rank both men among the smartest of players in their respective sports.

Many years ago, at a Yale faculty reception, one of us met President Ronald Reagan and talked with him for just a brief five minutes or so. Most of the faculty in attendance found him intellectually unremarkable and his politics, ideologically, quite distant from their own. However, he was blessed with a personality that put you instantly at ease, effortlessly casting a spell that created a pleasant illusion of relaxed familiarity that had no basis in fact. You liked him and he, quite convincingly, pretended to like you. President Reagan had remarkably high interpersonal intelligence, among the highest most will encounter. This mental ability, beyond the reach of an IQ test, shared by others such as Presidents Bill Clinton, Barack Obama, and, most assuredly, Franklin Roosevelt, is valuable and important in all societies. It has to be acknowledged that in addition to their high interpersonal intelligence, Presidents Clinton and Obama probably score highly on conventional IQ tests, too. Therefore, as we explore intelligence and view race through the prism of IQ, be aware that an awful lot that means a great deal is left out. While acknowledging their limitations, as we shall see, IQ tests do tell a great deal about some very important things.

Paradoxically, intelligence testing, now sometimes attacked as a tool for discriminatory sorting and invidious comparisons, grew out of a search for fairness. Their creation was prompted by a desire to avoid arbitrary decisions about which children would be allowed to benefit from enrollment in the French public schools. By 1900, French society was a passionately democratic one, devoted to providing free public education to all children. However, while universal public education worked for many, there were some children who seemed incapable of performing at levels that allowed them to keep up with their more proficient classmates. Rather than leaving the identification of such students to informal assessments of teachers and principals, the Ministry of Education decided to develop an objective instrument to perform these evaluations and hired Alfred Binet, a psychologist, to develop a procedure for doing so.

Realizing that at certain ages most children can do this or that mental task but at a younger age cannot, Binet came up with the notion of mental age. Calling on the experience of many teachers, he formulated lists of tasks that could be performed by most children of one age and those older but not by most who were younger. A child of five who could perform the tasks expected of a child of five was assigned a mental age (MA) of 5. If this boy could perform those expected of a two- or three-year-old but not those expected of a four-year-old, he was assigned an MA of 3. On the other hand, if a five-year-old could perform the tasks of a six- or seven-year-old but not those of an eight-year-old, she was assigned an MA of 7, and so on. Therefore, if the entry level for school was an MA of 5, Binet's system offered the prediction that a child of five years old with an MA of 5, or higher, would be likely to succeed, but a five-year-old with an MA of 4 would probably struggle and might fail, and a five-year-old with an MA of 3 would be very likely to fail.

Within a few years, psychologists began dividing MA by chronological age (CA) and multiplying the quotient by 100 to obtain what was dubbed an intelligence quotient (IQ); in short:

$$MA/CA \times 100 = IQ$$

Our five-year-old with an MA of 5 will have an IQ of 100. The majority of children five years old will have a mental age of 100 because such a mental age describes what most five-year-olds can do but four-year-olds cannot. However, the five-year-old child with the MA of 3 would be assigned an IQ of 60 and one with the MA of 7 has an IQ of 140. While a scale based on mental ages works for populations under sixteen years of age or so, it is unsuitable for adults—a thirty-year-old who can do what the average thirty-year-old can do has an IQ of 100 but, absurdly, an IQ of 133 if she can do the same things the average forty-year-old can do.

In the years since the development of the initial tests to measure the IQ of children, formulations suitable for adult populations have been developed. Batteries of questions, usually testing verbal, mathematical, logical, and perhaps spatial abilities are composed and given to large numbers of people. Questions everybody answers correctly and questions nobody answers correctly are dropped because they are not useful for differentially ranking performances on the test. While the exact procedures for formulating modern IQ tests are too elaborate for our purposes, the tests are composed in such a way that 50 percent of the population falls above and 50 percent falls below a score, known as the median score. That score is assigned a value of 100. When one examines the distribution of scores in a very large population of test takers chosen to represent all segments of the population for whom the test is intended, it forms a bell-shaped curve called a normal distribution. The mathematics of the normal distribution tells us a good deal about this bell curve of IQ. Fifty percent of the IQ scores in a population with a median of 100 will fall between 90 and 110, about 16 percent of scores will lie above 115 or below 85, and only about 1 percent of scores will be above 135 or below 65.

What is IQ? What do IQ scores tell us? "IQ is what the test measures" is a simplistic but accurate answer to the first question. Approaching the second with a thought experiment provides a perspective on the landscape of IQ. Imagine a group of one million people randomly chosen from the American population. Impose only the conditions they are over thirty years of age and born and raised in the United States. From this very large population, three groups, each consisting of 100 people, are selected. The only thing you are told about the members of these groups is their score on an IQ test. Members of Group 1 have an IQ of 80, members of Group 2 have an IQ of 100, and those of Group 3 scored 130. What can you infer about their level of education and the professional niches they might occupy?

While many traits—diligence, an ability to focus, industry, and so on—will be important to complete the long and intellectually demanding education and training required to earn a degree in computer engineering or to become a neurosurgeon, many people with an IQ of 130 would be able to perform the intellectual work required. However, all other things being equal, and they seldom are, not many of the individuals in the population that scored 100 would gain degrees in these areas, and it would be surprising if any of those scoring 80 would earn degrees in computer engineering or join the ranks of board-certified neurosurgeons.

We are not suggesting that everybody with an IQ of 130 is a computer engineer, only that those who are computer engineers are likely to have high IQs. Not every tall person is an NBA star, but NBA stars are very likely to be tall. There is nothing bold about asserting that IQ tells something about the likelihood of success in an educational program. After all, IQ tests were originally designed to do precisely that. Although acknowledging IQ as a predictor of trainability, we know that many other factors—commitment, work habits, background, and opportunity—are important factors, too. It is hard to imagine a university that would measure IQs and, based on the results, hand out or deny degrees in engineering.

WHAT DETERMINES IQ,
HEREDITY OR ENVIRONMENT?

Where does IQ come from? Do we mostly acquire it from our environment? Is it primarily encoded in our genome and, therefore, genetically determined? Is it some of both? Put another way, is it like height, is it like speaking Italian, or is it like singing Italian unusually well? A large body of evidence shows that stature is mostly determined by genes. If IQ is like height, then we need to focus on the genome and genetics. If you want to generate seven-foot-tall NBA candidates, the China strategy of finding very tall men and very tall women and persuading them to produce offspring has a reasonable prospect of success. The children are likely to be tall and, occasionally, remarkably tall. Therefore, being tall is strongly genetically determined. Speaking Italian seems to be very common among children born and raised in Italy, yet Italian families with very young children who immigrate to the United States tend to have children who generally speak English instead of their parents' more euphonious and beautiful language. Therefore, speaking Italian must be under a strong environmental influence.

Some of the most beautiful vocal music ever written is written in Italian, and although it is sung by many Italians, only a vanishingly small minority of Italians sing it remarkably well. In fact, among the few people in the entire world blessed with voices that allow them to sing Verdi or Puccini very well, the majority are not Italian. Moreover, even the small minority that can enchant with their operatic renditions were not born singing "Nessun dorma." They will have had the benefit of extensive training and countless hours of practice. However, unless they were born with the genetic endowment to develop an outstanding vocal apparatus and the accessory muscular and nervous systems to support it, no amount of training is going to make them Plácido Domingo or Anna Netrebko. We have to conclude that even though they are environmentally developed, great voices are products of heredity. What about IQ?

The short story is that both genes and environment contribute to IQ but not equally. Although this issue has been studied using a variety of approaches, in general, most indicate that genetics plays a greater role than environment. What is this evidence? To begin, the IQs of family members are more closely correlated with each other than with those of unrelated persons. The degree of correlation is quantitatively stated by the correlation coefficient. Recall that things that are perfectly correlated have a coefficient of 1 and those that are completely uncorrelated have a coefficient of 0. With regard to values in between, correlations of 0.40 to 0.70 are medium to high, and coefficients above 0.70 are quite high, while those around 0.30 and below are modest and imply even less of a relationship as they fall. In the case of parents and biological children living in the same household, a correlation of 0.42 has been reported. The correlation among siblings is 0.47, about the same. These correlations are significantly higher than the modest correlation of 0.19 found for the IQs of adopting parents and their adopted children. The IQs of adopted children correlate more closely with those of birth mothers than with the scores of adoptive mothers, 0.42 versus 0.24. In general, the IQs of unrelated adults raised apart show a correlation coefficient of 0.01 and are uncorrelated.

This data is certainly consistent with genes (heredity) playing a highly important role in the determination of IQ. However, one could argue that nuclear families living together share more than genes. They also share an environment. Because they live in similar environments, a good deal of the correlation of their IQ scores might be traced to environmental factors. However, the studies involving adopted and biological siblings raised within the same household showed little correlation in their IQs, despite living their lives in similar environments. This finding is consistent with genes making greater contributions to IQ than the environment. However, the most compelling evidence emerges from studies of twins.

We all know that twins come in two flavors, fraternal and identical. Genetically, fraternal twins, which may or may not be of the same sex, have 50 percent of their genes in common. When raised together in the same household, their IQs have been found to show a 0.60 correlation, significantly higher than the 0.47 reported for siblings that are not twins. However, in addition to sharing a mother as all siblings do, twins share a womb and day of birth. These environmental similarities distinguish them from non-twins. The finding that they have more similar IQs than siblings who are not twins (but still have 50 percent of their genes in common) is consistent with environment playing a role in shaping IQ.

Identical twins have essentially 100 percent of their genes in common. They are actually a small clone of humans. The IQs of identical twins reared together are very highly correlated, with coefficients of 0.86 reported, much more highly correlated than what is observed for fraternal twins who share only half as many genes. Such a high correlation is consistent with the genome playing a major role in the determination of IQ. However, most of us have experienced the striking and sometimes eerie physical similarities between identical twins. Individuals so similar to each other will be likely to inhabit environments that are more similar than those of non-identical siblings, even if those siblings are fraternal twins. If one twin is male, six-foot-two, good at mathematics, and athletic, his twin will also be male, probably also prefer sleeping on a king-size mattress, be smart, capable of playing a sport well, and, like his twin, have a good shot at admission to a highly competitive college. Perhaps the higher correlation of IQ scores is also strongly determined by the similarity of the environments that their similarity makes it likely they will inherit.

Studies of adopted children provide powerful insights into the relative roles of genetics and environment in the determination of traits such as IQ. A study of fraternal twins reared together as opposed to raised apart showed a correlation of 0.60 for those living together but only 0.38 when raised apart. This is in marked contrast to what is observed when the IQs of identical twins raised together are compared with those of identical twins raised apart. Several such studies have been performed, and they all had similar findings to the landmark study of Thomas Bouchard and his collaborators. They found that when identical twins were reared together, a correlation of 0.86 is observed for their IQs. Strikingly, the studies of the Bouchard group and those of other groups found that the correlation between IQs is only slightly lower at 0.75 when they are reared apart. This result strongly supports a major role for genes in the determination of IQ. However, even this dramatic study is not as clear-cut as one might assume. Though powerful, even adoption studies have their weaknesses.

For the purposes of determining the role of genes in determining IQ, children should be randomly assigned to adoptive families and homes. Fortunately for the children, adoption agencies do not use tables of random numbers to make assignments. Many factors, including socioeconomic factors, sometimes religion, sometimes ethnicity or race, almost always a combination of these and other factors, too, are factored into decisions about who will be adopted by whom. Certainly, these appropriate and humane practices result in children being adopted into environments that are more similar than would be the case if a random assignment was used. While the correlations for twins raised apart are quite high, some degree of environmental similarity probably makes a small but real contribution to the high levels of correlation observed.

DOES IQ DIFFER AMONG POPULATIONS, AND IF SO, WHAT IS ITS SOCIAL SIGNIFICANCE?

IQ differs among individuals but how about among populations? What would be found if one measured the IQs of different populations and compared them? The answer is that it depends on the populations compared. Certainly, differences among some populations have been found. For example, here in the United States, such testing has identified a population with a median IQ 15 points above another. These studies have also identified a population with a median score about 5–7 points above the population with the inferior median score mentioned previously. The three populations we have just compared are: Ashkenazi Jews, a group with a median score about 15 points higher than that of non-Ashkenazi Whites; East Asians (collectively Chinese, Japanese, and Korean Americans), a group that scores about 5 points higher than non-Jewish Whites; and the non-Jewish White population, which has the lowest median score of these three groups. Although Ashkenazi Jews (the majority of the American Jewish population is Ashkenazi) are identified and self-identify as White, there is a degree of genetic relatedness within this population and, for many, a significant set of cultural connections that are distinctive. In many cases, Ashkenazi Jewish ancestry can be determined by appropriate genomic analysis. However, it is important to point out that for many who fall under the umbrella term White—Italians, British Islanders, Norwegians, Greeks, Armenians, and many others—subgroups could also be discerned by genome sequencing. That said, a 15-point difference in median scores is a large one, and it is worth exploring the implications and apparent consequences of this finding.

Just for perspective, realize that the mathematics of the bell curve will put about 2.5 percent of the White population, which has a mean IQ of 100, above an IQ of 130. Having a higher mean IQ of 105, the East Asian American population will have 2.5 percent of its population above an IQ of 135, and the Ashkenazi Jewish population will have 2.5 percent of its population above 145, a striking number when one realizes that only 0.15 percent of the White population will have an IQ above 145. To the extent that IQ influences intellectual performance, we would predict a disparity in the intellectual achievements of these groups. When we compare Ashkenazi Whites with the non-Ashkenazi White population, disparities are apparent and unequivocal. Disproportionate numbers of Nobel Prizes, Pulitzer Prizes, and MacArthur Awards have been won by American Jews. Although Jewish Americans are only about 2 percent of the population, a little over a third of the Nobel Prizes won by Americans have been awarded to members of this group. With regard to such indices as scholastic achievement, particularly in mathematics, East Asians also outperform the White population. By more global measures of success, such as family income, both Jews and East Asians also outperform the White population. Despite these facts, none of which is in doubt, the underperformance of the White population on IQ tests is not a cause of great concern.

In general, the fact that non-Jewish Whites finish last in an IQ derby when compared with Jews and East Asians is hardly viewed as their most pressing problem by Whites lunching at the Harvard Club or deciding who to admit to the membership at the Augusta National Golf Club or to other enclaves of great power and wealth. Nor is it a concern of the Whites populating the pages of *Hillbilly Elegy*, J. D. Vance's revealing memoir of life lived by a quite different group of Whites in Kentucky's Appalachia and in Rust Belt Ohio. Concern is much higher about the finding that life expectancy among middle-aged, working-class Whites has begun to fall. Increased rates of suicide, increases in alcohol abuse, and the invasion of some rural and working class White communities by such socially disruptive and destructive opioids as heroin are thought to contribute to this decline. The opioid epidemic, an unexpected and additional burden to some once vibrant and stable places, is regarded by most Whites, and others, too, as a much more pressing concern than the nonproblem of intergroup differences in IQ scores.

Two or three generations ago, many of these places were economically vital centers of manufacturing and mining. While out-migration and college were pursued by some, most exercised the once viable option to stay at home with a good and stable job, often with good salaries and welfare benefits negotiated by strong labor unions at the local mine, mill, or factory. Then things changed, and what had been a successful and winning strategy became a losing one. While increasing numbers of Rust Belt, rural, and Appalachian Whites find themselves defeated by the challenges of life in a changing America, it does not occur to them or to serious students of the situation to blame the defeats of Whites like John Steinbeck's Joad family of yesteryear or Vance's Whites of today on the tyranny of a bell curve. For understanding, we must look to a diversity of systemic factors.

What we see are shifting economic forces, such as the movement from coal to natural gas, wind, and solar power; the decline of unions following the loss of the smokestack, extractive, and factory cultures that spawned them; the arrival of disruptive technologies that replace people with robots and displace common sense with artificial intelligence; the burdens of the economic tax of poor health care and the poor health caused by the lack of an umbrella of health insurance; and the entrapment of too many in outmoded or maladaptive cultural patterns ill-suited or poorly aligned with integration into a productive social fabric.

Despite these difficulties experienced by a significant subpopulation of American Whites, they are not perceived to have and, in fact, do not have a consequential achievement deficit. The many scientific, technological, and industrial landmarks produced by America's White population—James Watson's co-discovery of the structure of DNA, arguably the master key for unlocking the secret of life; Thomas Edison's many inventions; the industrial colossus assembled by John D. Rockefeller—are a few emblematic examples. The majority of the world's best colleges and greatest universities were established by this group. This group has given the world the iPhone, the personal computer, and Microsoft (one its useful programs wrote and spell-checked this sentence). The White entrepreneurs who founded these companies have been happy to accept mountains of the world's cash as thanks. The reality of so many achievements in so many areas makes a discussion of intrinsic White intellectual inferiority short, silly, and of little import, allowing the dismissal of the documented White IQ deficit as a matter for concern.

On the other hand, this overview of their accomplishments—many, diverse, and major—must be balanced by a reminder that this population has a history of moral underachievement. It has been slow to see its mistakes and has required repeated episodes of special education to repair ethical deficits, needing do-over after do-over. And still, in too many cases, they fail or refuse to make the right and necessary corrections. An accurate recital of specific failings would be long and would surely include the ugly stains of White America's genocidal oppression of American Indians and the appropriation of their land. It would include the oppression of many waves of immigrant groups, including East Asians, White subgroups, of which Jews come first to mind, but there are others. A segment of this population enthusiastically supports a continuing campaign against those from Mexico eager to contribute their work and talent to the American economy. And, of course, there was the blight of Black slavery and the long and still continuing aftermath of racial animosity and injustice.

THE REAL CONTROVERSY IS OVER BLACK VS. WHITE IQ DIFFERENCES AND WHAT THEY MEAN

In contrast to mild interest in mean IQ differences among the groups just discussed, we have become accustomed to a strong and persistent focus on IQ differences between Black and White populations. Here, too, the reality of the difference is a matter of fact. Randomly select a thousand American Whites and Blacks and administer IQ tests, and the Black population will produce a mean IQ of around 85 and the White population a median of 100. There is a strong consensus on a 15 point difference in median IQs between these two populations. Argument about these differences distills to a potent version of the debate over the accuracy of President Jefferson's simple and profound declaration that all men are created equal. Although his statement of this principle was eloquent and unequivocal, President Jefferson, a brilliant and complex man who held slaves and a conviction of their inferiority, did not believe Blacks were the equals of Whites. He is not alone. Our American history is convincing in its documentation of the difficulty we have had and continue to have in fully embracing Jefferson's proposition of a declaration of universal equality. With regard to IQ, the argument is primarily over this question: Is the large difference in intergroup IQ environmentally or genetically determined? The question does have implications for whether one should expect equality of opportunity to eventually produce equality of outcome. This is a difficult question to address empirically since deep equality of opportunity has not yet been achieved.

The foundational manifesto of the quantitative quarrel is a data-based, carefully reasoned, and skillfully argued article in the *Harvard Educational Review*, published fifty years ago by the distinguished behavioral psychologist Arthur Jensen. The core conclusion of this influential article was previewed in its title: "How Much Can We Boost IQ and Scholastic Achievement?" Jensen and his successors begin with the reasonable premise that IQ is the best general predictor of academic achievement. Therefore, populations that have higher mean IQs will have greater success in school than those with lower scores. They go on to argue that although genes and environmental factors play a role in determining IQ, genes are the major actors. Since environments can be changed, it is possible to modify environmental effects on IQ. However, for the purposes of this discussion, an individual's genetic makeup is not subject to manipulation or modification. This conclusion aligns with the twin and adoption studies reviewed earlier. The next step in the argument is to state that the major factor determining the difference in the IQ scores of different populations is genetic, determined by DNA, with the environment making a real but minor contribution. The difference in the mean IQ scores of Black and White populations predicts that the White population will perform better in school than the Black population. Since the major cause of these group differences is genetic, social interventions to improve the environment, if equally available to Whites and Blacks, will not erase the difference because they will not change genetic makeup; that factor, they assert, is responsible for most of the difference between these groups. Consequently, a large investment of funds and social energies in attempts to eliminate or significantly narrow this gap by environmental interventions will fail. That, in somewhat simplified form, is the core of the hereditation position. We don't know whether it is correct or not.

We do know that subscription to either the hereditation or environmental interventionist position is likely to result in vastly different social policies. Thoughtful and situationally appropriate environmental interventions on a massive and national scale over a period of a couple of generations would be unlikely to produce a society less capable than the conspicuously stratified one we have now. Indeed, it would be interesting to see the outcome of such a social experiment. However, it has yet to be pervasively conducted. In the mid-1960s, a start was made when the Great Society programs of the Johnson Administration were instituted. However, their vigorous pursuit and hopes for expansion were curtailed with the arrival of the Nixon Administration and President Nixon's determined and successful efforts to take the country in a direction quite different from that of his predecessor.

If such programs of educational enrichment were started again, their goal should not be to narrow the Black–White IQ gap. They should be to raise levels of scholastic achievement in groups where there are deficits. Such programs would target Native Americans, some Hispanic populations, lower-income Blacks, and some White populations as well. Perhaps in addition to providing a better, richer educational experience and probably raising the scholastic achievement of all these populations, such a social investment might pay the additional dividend of narrowing the achievement gap between the White Ashkenazi and some non-Ashkenazi White populations, too.

While IQ gaps will surely continue to interest those whose livelihood depends on their study, what will be important to the larger society is a better-educated, higher-achieving, and more capable populace.

IN SUMMARY

Given the peculiar history of the United States, a suspicion of Black intellectual inferiority will remain despite assertions to the contrary. This is not a notion that will be argued away. It's like sports. Initially, Jackie Robinson's success on the diamond surprised people—"Well, at least *one* can." When he was joined first by a few and then by many outstanding Black players, perceptions changed, and people forgot there had ever been a question about the ability of Blacks to play major league baseball. In football, for many years, the quarterback position, which required field generalship, judgment, and great poise under pressure, tacitly excluded Blacks. Quarterbacks had to be smart. Something beyond mere "athleticism" was required. Then there were the "first" successful Black quarterbacks in major college football followed by the National Football League. Subsequently, the numbers have grown large enough that the ability of Blacks to tenant this key position has become an established fact and no longer a question. There was a shift in perception from "Can they do it?" to "Of course they can do it because they are doing it." People will abandon their notions of Black intellectual inferiority when their experience shows them there is nothing unusual about being smart while being Black.

Increasingly, we see Black achievement in many areas becoming quotidian and unremarkable. It is no longer exotic. For many years, Black contributions to literature—numerous, varied, and conspicuous—have challenged the very low expectations encouraged by the bell curve. The past twenty years or so have seen the ranks of sharp, savvy politicians and journalists swell with an influx of Blacks clearly able to hold their own on the page or in front of the camera. Black faculty who have earned a place as peers of their colleagues have become commonplace on many campuses and in many disciplines. It has become routine to see Black faces populating the ranks of commentators and public intellectuals appearing on platforms, panels, and on PBS. Talking heads have become nappier while remaining thoughtful and articulate. At last, our society is finding the content of their comments more interesting than the color of their skin.

Black achievement in politics, literature, the arts, and, increasingly, the social sciences continues to grow and to impress. When was the last time we wondered if Blacks might have the intellectual horsepower to write a good novel? In 2008, whether a Black person could run the world's richest and most powerful country was an open question. After eight years of accomplishment, carried off with admirable grace and consistently deploying high levels of several kinds of intelligence, we now know that at least one can, and we suspect that other Blacks, including his remarkable wife, could, too. However, there is still much to be done. Black scientists, mathematicians, or engineers continue to find themselves lonely exemplars of participation and achievement in these key disciplines. While Neil deGrasse Tyson informatively and entertainingly demonstrates that a Black man can understand and even help others understand the most fascinating and arcane facets of astrophysics, he is still one of too few waiting, hopefully yearning, to be joined by many more Black men and women.

Chapter Six

Seeking Solutions

"All animals are equal, but some animals are more equal than others."
—from *Animal Farm* by George Orwell

"It's not easy bein' green."
—Kermit the Frog

Difference is a problem that racial minorities, ethnic and religious ones, too, have to confront. "It is not easy bein' green" might be Kermit the Frog's most perceptive observation. For many years, diversity has been a hallmark of the US population, and it has often been a divider, fragmenting us into camps, ignoring what we have in common and focusing on our differences. When those differences are between a powerful majority and a weaker minority, there is the potential for oppression. This potential is realized when majorities deny equality to racial and ethnic minorities. Despite the guarantees made in the 14th Amendment to the US Constitution, examples of failure to extend equal protection by the law to all citizens are readily found. Compromise or denial of equal access to the economic, educational, and social resources that facilitate success to members of some racial and ethnic groups has been a familiar and persistent theme in American history. Any humane solution to America's race problem will have to confront inequality and replace it with equality.

Race and the problems it has generated are among the oldest, most durable, and most defining features of our cultural and political landscape. Before the founding of the Republic, exploitative confrontation with Native Americans and the importation of African slaves established racial themes that would generate factions and, eventually, cause wars. Racial mythology and polarization have lured the ruling majority into the commission of moral outrages that have indelibly stained our national conscience. The past, present, and future contours of our political landscape can be seen and understood only when viewed against the complex and troubling backdrop of race. The interaction of Whites with Native Americans and with enslaved African populations and their descendants are the two major features of racial history in America.

TREATMENT OF THE NATIVE AMERICANS—ONE OF AMERICA'S GREATEST MORAL FAILURES

In one of our country's greatest moral failures, a partial genocide of Native Americans was the approach employed by Whites to solve the race problem between these groups. This genocide was an amalgam of direct and indirect approaches. The direct approach involved war and deliberate economic disruption. Indirect avenues involved the transmission of diseases endemic to Europe and Africa to New World populations. These indigenous populations had little immunity because their immune systems had not been trained by prior exposure to pathogens, unwittingly imported by European Whites and the African slave populations they maintained. For example, while smallpox, endemic in Eurasia and Africa, was a feared and serious disease in the Old World, it was unknown in the Americas before the arrival of Europeans. Introduction of smallpox (at first unintended, but later deliberate) was devastating in immunologically naïve Native American populations. Some scholars, most notably, Jared Diamond in *Guns, Germs, and Steel*, identify the impact of smallpox and other infectious agents imported from the Old World and spread among the indigenous populations of the Americas as the decisive factor in the conquest of the New World. While their arms, including guns and horse cavalry, were superior to those of native populations, the small numbers of Spanish conquistadors were insufficient to subdue and maintain dominance over indigenous populations. In aggregate, these native populations numbered in the tens of millions and were broadly settled over millions of acres on two continents. Furthermore, the Americas lay across a wide and challenging ocean, far from the European bases of aspiring conquerors, which rendered their massive reinforcement impossible. As Diamond succinctly concluded, it was germs, not guns and steel, that stole the Americas.

While a mostly unwitting germ warfare was the major ally of conquering Europeans, here in the United States, destruction of their economic life was a key feature in the partial genocide of the Native American population. In addition to the attrition of raids and wars, forced relocations took an enormous toll. Tribes were moved from regions where they had lived for many generations, built distinctive cultures, and worked out effective adaptations. They were forced to settle and begin anew in faraway regions with different climates, plants, and animals. It is no surprise that these relocations took an enormous toll on lives and indigenous economies.

Thus ecological, as well as conventional, war was waged on native peoples. For example, the economy of some Indian tribes of the Great Plains was centered on the buffalo. Their meat was food. Their hides were clothing, footwear, and blankets. Tools were fashioned from buffalo bones, and dried buffalo dung provided fuel for fires that cooked and warmed. These economies were severely disrupted by the deliberate slaughter of millions of these creatures. Destroying the buffalo population destroyed a people's way of life, facilitating their conquest and liberating land for appropriation by an expanding and imperial young America. This strategy and others discussed previously reduced North America's Native American population, once in the tens of millions, to scattered remnants. In the years since, many descendants of Native Americans have become part of the White population by intermarriage; others, intermarried or not, have continued to identify as Native American; and still, some continue as members of one of the more than 560 remaining Indian tribes. Today, the tribes, which are both cultural and political units, have a complicated relationship with the US government and with an American population that is widely remorseful, embarrassed, and, finally, apologetic.

SLAVERY'S ENDURING STAIN

Black slavery and its aftermath have generated a quite different set of race problems. Unlike Native Americans, Blacks were a part of the machinery of nation-building, not a contending force to be subdued. Although Blacks have been essential partners in American economic development, the partnership was founded on inequality. Beginning as a master-slave relationship 400 years ago in 1619 when slaves were brought to what would become the United States, Blacks were slaves in the Americas for almost two and a half centuries before becoming a free people for the century and a half since the end of the Civil War in 1865. During the first two-thirds of the relationship, most Blacks, treated as work animals that were bought and sold, were in a profoundly unequal position, having little more control over their lives than other livestock. Despite the goals of Reconstruction, for 100 years or so after the Civil War, Whites maintained or developed legal frameworks and social structures, such as Jim Crow, that preserved and reinforced economic and social inequality between the two races.

The central problem in Black–White race relations has been and continues to be imposed and constructed inequality. Beginning with the 13th, 14th, and 15th Amendments passed during Reconstruction, followed by landmarks such as *Brown v. Board of Education* and the civil rights legislation of the

Johnson era, great strides have been made in effecting de jure solutions to the central problems of race in America. There is now a legislative shield protecting Blacks, as well as others, against a broad variety of depredations. The legal framework is firmly in place to prevent racial discrimination and race-based inequality. However, when one looks at the de facto situation—the degree to which there is actual equality of opportunity, equal access, and equal treatment; the extent to which systemic factors that intentionally or unintentionally promote inequalities between Blacks and Whites have been eliminated—one finds our society still seeking solutions. Economic inequalities and the unequal operation of the justice system and its agents are major barriers that persist, stubbornly blocking progress. These will have to be gotten around, or better yet removed, before the path to solving race problems in America can be negotiated successfully.

MODELS FOR MINORITIES

Confronted with the difficulties of living within a society where a majority has often made life difficult for groups who are different, minorities need to develop strategies for effectively coping with the many challenges posed by the asymmetries of majority/minority relationships. Within the constraints imposed by the dominant population, different groups have approached this problem in different ways. It is informative and useful to examine the strategy and tactics of a particularly successful minority and identify some of the elements of an approach that has enabled it to flourish in spite of cultural difference and overt discrimination.

By any one of a variety of measures, despite the obstacles of a once-virulent anti-Semitism, the American Jewish community is perhaps the most broadly successful minority in the United States. Although, as we will see later, the ancestors of most of America's Jewish population arrived poor and with very little education, today this population is largely well-educated and has a high median income. Jews are conspicuously well-represented in the professions, the academic sector, and finance, and several members of this population play important roles in shaping a variety of national policies. Though America's Jewish population is now widely and deeply socially integrated into the larger fabric of society, it still manages to maintain a self-affirming identity as a distinct ethnic group. The experience of one of us offers insight into how some members of this minority managed their relationship with the larger culture.

Confronting the challenges of a distinctive minority living within the culture of a large and dominant majority, a group of Jewish parents coalesced around a shared concern for the options open to their children. The families were prosperous and lived in New York City, with a substantial Jewish population that came primarily from European countries, and they hoped that their children would have genuinely equal access to the colleges and universities across the land. They were aware that many institutions had established secret quotas for Jewish students and hoped that their children would grow up able to move fluently in gentile society while remaining rooted in their own religious and cultural traditions and valuing these differences. They sought intellectual excellence and knew that inclusiveness with high quality might be harder to achieve for girls than for boys, since they historically had not emphasized it. Fortunately, they had the resources and imagination to achieve what they desired. They found it in The Brearley, a single-sex (all female) school that developed relationships with several other single-sex (all male) schools in the gentile world.

At first glance, it was a rather traditional institution: students wore uniforms, classes were formally structured, and weekly assemblies began and ended with hymns and readings from the Christian tradition, but the emphasis was ethical rather than religious. History was taught with an emphasis on the cultures of different nations and eras, so that, for example, in the year when students studied "the ancient world," in addition to studies of Greece, a significant amount of time was spent on ancient Israel. There was significant study of the rise of Christianity, perhaps more than some of their Jewish contemporaries might have appreciated. All this was accompanied by ancient texts and artwork to bring it to life as much as possible. It has often been said that "you can't appreciate English literature unless you understand the religious context," but the school made it clear that appreciation did not require affiliation. Diversity of other kinds was also sought and continues to be sought. Notably, the school was ahead of most of the wealthy private schools in the city in admitting Black students, and during the years following World War II, when considerable numbers of refugees and displaced persons ended up in the city, it provided scholarship aid and language tutoring for promising applicants.

It was from this model that one of us acquired the image of how a society might include a persistent minority community, not requiring assimilation but appreciating their participation and contribution to a diverse whole. It is true that a sophisticated and flexible member of contemporary American urban society needs to be moderately familiar with various forms of Christianity and Judaism (and other traditions also) and to be sensitive to many of the ethical issues involved but can still be deeply committed to a different set of values and traditions, in this example, Jewish values and traditions. The "melting pot" is not the only solution—every spoonful can be a pleasing surprise if the cook sees each ingredient as offering something of distinctive value to the whole without being reduced to mush. No American should be forced to deny or be cut off from his or her origins. Indeed, among the horrors of the slave trade was the deliberate practice of putting captives into environments where none of the other enslaved Africans came from the same tribal background or shared a language. Both similarity and the complementarity of difference can contribute to strong and productive relationships between persons and between peoples.

MINORITY MANIPULATORS OF THE MAJORITY—BOOKER T. WASHINGTON AND MARTIN LUTHER KING JR.

One of the important lessons that should be learned from the example just explored is that minorities are wise to thoughtfully manage the relationship with the other communities with whom they live. One of the most useful tools minorities can employ is an in-depth knowledge and clear-eyed understanding of those communities. The history of the Black struggle for civil rights has from time to time demonstrated a deep and realistic understanding of the White population, and it has fared best when it has used those insights to manipulate a dominant majority population in ways that made the best of difficult situations. In what follows, consider two very different examples of positive social manipulation based on accurate and sophisticated readings of the targeted population.

After the Civil War, toward the end of Reconstruction, throughout the South, there was a concerted and determined effort to remove Blacks from political positions they had gained during the postwar occupation by Union forces. This effort, often brutal in its conduct, had the goal of returning former slaves to positions at the bottom of the social hierarchy, providing cheap labor for menial tasks. Many Whites in the South, with the tacit agreement of Northern White allies, viewed this as the natural order of things. Although embedded in this climate, Booker T. Washington managed to play a central role in building the Tuskegee Normal and Industrial Institute. It was situated in a region of Alabama where there was a firm determination in the White population to keep Blacks "in their place" on the lowest rungs of the economic and social ladder. At that time and in that place, there would have been little tolerance for a Black ivory tower, a bookish enclave for Blacks to immerse themselves in classical study.

Acutely aware of the likely resistance to providing a traditional college education to Blacks, Washington framed his ambition in terms that would not provoke the responses, ranging from economic reprisal to Ku Klux Klan violence, reactions that would defeat his efforts. Instead, he encouraged the notion that the intention of the Tuskegee Institute was to train Blacks so that they could be of greater and more skilled service to the White community. Tuskegee would train plumbers, not poets, and graduate masons, rather than mathematicians. It would also train teachers who would go forth from the Tuskegee Institute and raise the practical competence of Blacks in other places, swelling the ranks of those with the skilled capacity to serve. Washington used his oratorical gifts to placate and reassure, sprinkling his talks with platitudes such as "do a common thing in an uncommon way" and "there is as much dignity in tilling a field as in writing a poem." To allay White fears that Black economic progress would give them ambitions to join Whites socially, in a widely quoted speech, he stated: "In all things that are purely social we can be as separate as the fingers, yet one as the hand in all things essential to mutual progress."

For its time, Washington's approach was impressively successful. The Tuskegee Institute he founded had around 100 acres and no endowment. Thirty-four years later, at the time of Washington's death, it was over 2,300 acres and boasted an endowment of a little over 1.5 million dollars, the equivalent of about 37.5 million dollars today (in 2014, the median endowment of private colleges and universities in the United States was 7.9 million). Washington died in 1915, as much an American icon as it was possible for a Black man of his time to be.

Though successful and pragmatic, his accommodationist approach was deeply flawed by its surrender of equality, conceding a hierarchy in which Black training and aspiration would be subservient to and shaped by the needs and tolerance of White society. Such a servile approach was calculated to gain the acquiescence, even the approval, of many Whites who were well-served and not threatened by the training of a color-coded helper caste. However, this approach would not earn the respect necessary to gain recognition as equals. In *Souls of Black Folk*, W. E. B. Du Bois levels a devastating critique of Washington's embrace of limited intellectual and social horizons for Black people, arguing that education and training of Blacks would have to include philosophy as well as plumbing. While recognizing that it was good, even desirable, that some receive vocational training, Du Bois argued the necessity for some Blacks, "the talented tenth," to acquire higher education. He goes on to point out that intellectual development would go hand in hand with the vote in securing and wielding political power and influence. His were the first systematic arguments for solutions to the race problem that could lead to equality with Whites.

It was in 1956 that a young Baptist minister deployed a strategy of social manipulation that would be more effective in shifting the American moral compass than any previous one. Armed with the tactics of nonviolent civil disobedience and blessed with the compelling rhetorical and soaring oratorical gifts necessary to attract and hold national attention, the Reverend Dr. Martin Luther King Jr., more than any other Black leader, forced examination of the morality of one race's systematic denial of equality to another. Dr. King's strategy involved engineering highly provocative but peaceful confrontations with state, county, and municipal governments. These power structures of the White population were willing to use the legally sanctioned violence of their troopers and police to enforce a system of racial inequality. It was understood by King and became apparent to the nation that there was an eagerness in some of these institutional elements to commit acts of violence, some of them lethal, against Blacks and those Whites who aided their struggle. King understood that the American population viewed its society as one that was just and fair, not one that would condone violently attacking unarmed and peaceful men and certainly not vulnerable and defenseless women and children. He made the correct moral calculation that there was a bedrock of decency in the majority White population that could be mined and used.

Dr. King's strategy was a cold-blooded one that followed this nonviolent battle plan: deliberately arrange protest after protest by Blacks and others sympathetic to their cause against discriminatory policies and practices. Place these groups of clearly nonviolent protesters, often in a posture of prayer or marching along singing hymns to brotherhood, in situations that would provoke a confrontation with police. Whenever possible, arrange for these groups of protesters to include men, women, sometimes children, and, when possible, some Whites as well. Dr. King and his associates always made sure these intentionally provocative demonstrations were announced in advance and well-publicized.

As intended, many of the demonstrations induced brutal attacks by angry truncheon-swinging police and sometimes the attacks also included the use of fire hoses. There were even instances when K9 corps dogs were deployed and released to maul demonstrators. These tactics, nonviolently provoking police into assaulting peaceful and unarmed demonstrators, were ruthlessly and repeatedly employed by Dr. King and others. The sight of little children being set upon by dogs on the streets of Birmingham, Alabama, sickened Americans, and the attack on peaceful voting rights marchers at a bridge site in Selma, Alabama, shocked a nation. These were the effects on public opinion that were intended. As the struggle ground on, it became apparent that the segregationists were losing the public relations battle, and as the smoke cleared, King's movement occupied the high ground. However, victory came at a great cost. Casualties included many who were injured and some, eventually including Dr. King, lost their lives. Nevertheless, the seeds were sown for the harvest of civil rights legislation that would extend de jure equality to Black Americans.

EQUALITY AND INEQUALITY

We see a denial of equality, whether de jure or de facto, as fundamental to sustaining the race problem and believe that a solution will come only when full equality of opportunity is enjoyed by all groups. A fanciful approach to achieving this goal would be to issue a proclamation announcing a date for Equality Day, after which all laws and all policies would actually treat everyone equally, with no one being held back by discrimination and no one advanced except by individual objective merit. Though straightforward and appealing on first glance, reflection reveals serious flaws in this utopian approach. Treating everyone exactly alike does not necessarily produce equality. In fact, it could be unfair. Such a stance ignores the reality that some, through no action of their own but merely because of birth and circumstance, begin the race ten yards ahead of the starting line and others have to start ten yards behind it. Mindful of George Orwell's wry saying of "Some animals are more equal than others," we have to face the reality that by happenstance of birth into a particular social class, these children will grow up with predictable advantages but those will be mired in disadvantage.

Children born into comfortable middle-class or wealthy families will be likely to get better education and better health care than the average American. Truely developmental and enriching preschooling will have been part of their experience. They will live in secure neighborhoods. In contrast, children born into low-income families will attend schools that are less able to provide strong educational input and the social support they need. Instead of a structured preschool experience, some of the luckier ones may get daycare, but more might be babysat by the TV. Many of their health-care needs will not be addressed, in many cases because of a lack of money to pay for medical care. Other needs will also go unaddressed because they are unaffordable or, for various reasons, unrecognized by their parents. Too many of these children will live in neighborhoods that are less secure than those of middle-class Americans.

If social policies stop at simple equality—treat everyone exactly alike—having failed to provide equality of preparation, we will fail to provide true equality of opportunity. We will have encouraged, and should expect, a disparity of outcomes. Some of those born into better circumstances will roll into good lives pushed in by family and circumstance, rather than pulling themselves in by their own efforts. It is true that a few born into disadvantage but with exceptional drive and ability and blessed with good luck may achieve at levels that match those who grew up under more fortunate circumstances. However, many with drive and ability equal to those of privileged Americans but born into circumstances of poverty and disadvantage will not get the education or enjoy the social stability that equips them for equal competition with their peers who did not suffer disadvantage.

Everyone knows that history has been different for Native Americans and Blacks in ways that have consequences that reach all the way up to today. For a complex of reasons that differ between Blacks and Native Americans, a larger fraction of each of these groups than others populate the lower-income ranks. Many low-income children born into these groups will start the race from a position behind, even sometimes far behind, those with more advantaged circumstances of birth. Although to some degree influenced by individual behavior, a strong argument can be made for structural factors being influential determinants of higher levels of poverty in some groups than others.

In *Educated*, Tara Westover tells a riveting and often searing story of growing up with little money in culturally isolated circumstances in a family with many attitudes—especially those ignoring or dismissing the importance of systematic education and training—that are ill-suited to attaining educational and economic success. The story told in *Educated* is remarkable and surprising because, in spite of these handicaps, many members of the Westover family are conspicuously successful. Their success has an aura of the miraculous because of our quite reasonable expectations that a collection of disadvantages, such as those arrayed against the Westover family, would make an escape from an existence of narrow horizons and poverty unlikely. Rather than a bright hymn to overcoming adversity, *Educated* is a dark tale of narrow and improbable escape. On the other hand, reading the *Grapes of Wrath*, we expect that the structural conditions of the Great Depression and the precarious circumstances of migrant farm laborers will conspire to doom John Steinbeck's Joad family to a meager, precarious, and sometimes desperate existence. They do. They almost always do, and social policies should be inspired and drawn up with the usual, rather than the exceptional, as models.

BLACK VS. WHITE DIFFERENCES IN NET WORTH ARE MAJOR DRIVERS OF RACIAL DISPARITIES

Net worth is a very good indicator of advantaged or disadvantaged circumstances. In 2016, the median income of Black households was just over 60 percent of White household income. A likely consequence of this difference is apparent. Even if they practiced habits of much greater thrift than Whites, the lower incomes earned by Blacks, and other factors that will be explored, make it difficult for the Black population to accumulate median levels of net worth that approach, let alone match, those of Whites. A look at figures for 2016 shows that Black households had a median net worth of $17,600, nearly ten times lower than the $171,000 median net worth of White households.

Many consequences flow from differences in net worth. People with more net worth will find it easier to obtain mortgages and will be able to borrow more. This means that they will be able to buy houses in more desirable neighborhoods than people with lower net worth. Lower crime rates and lower levels of gang activity are likely to be among the features that make these neighborhoods desirable ones. Often, people living in more expensive neighborhoods have access to better education with more programs during and after school than those residing in poorer ones. Because their children receive better high school educations, it is more likely that they will pursue higher education and more likely that they will successfully navigate the college experience and complete degree programs. Although college expenses are high, families with higher net worth find it easier to finance college costs. They are more likely to have savings and, if necessary, they can borrow. One may find the police less antagonistic toward those they perceive to be residents of higher income areas than toward residents of economically depressed ones. Provided they are perceived to "belong," residents of neighborhoods populated by higher net worth households are likely to find their encounters with local police to be more cordial and less adversarial than those experienced by residents of economically depressed neighborhoods.

Furthermore, if there are confrontations with the law enforcement system, the ability to hire lawyers, pay bail, and bear the other costs of mounting a defense is dependent on financial resources. Consequently, higher net worth individuals have a better chance of getting charges dropped, winning acquittal, or, if convicted, getting fines and probation instead of incarceration. Shoplifting in a convenience store or prostitution on the street are crimes more often committed by those of low net worth than by those of high net worth. These offenses and others like them will land their perpetrators, mostly of low net worth, in jail. On the other hand, moral offenses that are far more serious and likely to wound many more victims much more deeply, such as schemes to get the elderly to make bad investments or enter into agreements that endanger their home ownership, are legally perpetrated by some higher net worth members of the financial sector. Unlike the street crimes mentioned previously, these moral crimes don't bring jail time; indeed, they make their perpetrators richer, ironically bringing the privileges, opportunities, and comforts of even greater net worth.

Although many factors go into determining net worth, for most Americans, their home is the largest component of their net worth. Sales of houses are rarely cash transactions in which the buyer forks over the cost of the house at the time of sale. Most Americans make an application for a mortgage to buy the house they have chosen, and then spend the next fifteen to thirty years paying off the mortgage in monthly payments. Like rent, the monthly mortgage payment provides a place to live. Unlike rent, each payment is a form of automatic forced saving with payments building increasing equity and, ultimately, discharging the debt, giving the buyer full home ownership. Moreover, there is another benefit to homebuyers. During the time of ownership, many houses not only hold their value but increase it. If a financial institution thinks the house has value, the amount of equity—reckoned on the basis of what has been paid toward the original mortgage debt plus whatever appreciation the house has enjoyed since its purchase—may serve as collateral for further loans. These loans can be used for the repair or improvement of the home, allowing the owner to preserve or even improve its value.

Loans can also be used for a variety of other purposes—securing medical care the family might not otherwise be able to afford, paying for a college education, or securing vocational training that leads to a job as a plumber or electrician. The children of home-owning families often gain additional wealth in midlife by inheritance when they receive ownership of their parent's house. Homeownership is a basic, versatile, powerful, and intergenerational financial tool. Any group denied full access to it in the past probably carries a handicap now.

BLACK NET WORTH HAS BEEN DELIBERATELY DEPRESSED BY FEDERAL LAW AND POLICY

Levittown, Pennsylvania, provides a revealing story of how exclusionary housing policies produced multigenerational effects on net worth. The end of World War II brought hundreds of thousands of veterans back to civilian life. Many of them were young and just starting families. Their needs, swelled by those of other house-hungry Americans, generated a demand for housing that greatly exceeded the supply. Perhaps inspired by the staggering feats of aircraft, tank, and ship production achieved by American industry during that war, William Levitt devised a system for the mass production of housing. At peak production, his organization was finishing a house every thirty minutes. On an area of twenty-two square miles near Philadelphia, he and his associates generated more than 17,000 houses to create a development eponymously called Levittown.

Levittown was an orderly and thoughtfully planned community of curving streets, with several schools appropriately distributed throughout the development to function as neighborhood schools. Levittown houses sold for between $7,000 and $9,000 and could be purchased with a down payment of 5 percent. Purchasers of the least expensive model could make a down payment as small as $350 and get a thirty-year mortgage with payments as low as $60 a month. For perspective, since $1 during the mid-1950s is the equivalent of about $9 today and translating yesterday's dollars to today's equivalents, you would pay $3,150 down and, with mortgage payments of only a little more than $560 per month, get a surprisingly well made house for only $63,000. Returning to the 1950s, gasoline was 25 cents per gallon, pot roast went for 45 cents per pound, eggs were 60 cents per dozen, and the average price of a house outside Levittown was a little more than $9,000. With costs at these prevailing rates, paying $7,000 for a two-bedroom Levittown house that came with its own nice little lawn was a very good deal for a family just starting out and making the average annual income, which was about $4,200 at the time.

And by the way, the retired offspring of the steelworker who bought that house back in the 1950s could sell it for somewhere around $170,000 today. Notice the heir, now a senior citizen, reaping a profit of $163,000 from the house's more than twenty-four-fold appreciation. All of this plus the benefits of two generations living in the house, enjoying life in a stable, safe community with decent schools and inviting playgrounds. "But wait, there's more!" as they bellow on late-night TV. The homeowner also received the small benefit of using a deduction of mortgage interest to reduce state and federal income taxes. Also, when the house was sold, the profit collected by the steelworker's heir was taxed as a capital gain, 0 percent for incomes less than $39,000 and 15 percent for incomes less than $434,000 in 2019. For most taxpayers, this is a significantly lower rate than that paid on the earned income that comes as a paycheck or fee for services rendered. However, as we will see, federal law denied Blacks their Levittown story.

Obtaining the huge financing required to build this remarkable place called Levittown was challenging. It was the Federal Housing Administration (FHA) that enabled this project by guaranteeing the loans necessary to carry it out. The Veteran's Administration (VA) also played an important role in facilitating purchases in Levittown as it did all across the country by providing low-interest, long-term loans to veterans for the purchase of houses. However, both the FHA and VA, agencies of the US government, publically decreed that none of the Levittown houses could be sold or rented to Blacks (or Negroes, in the parlance of the day). Furthermore, when you bought your Levittown house, like all buyers, you were required to sign and honor a "restrictive covenant," a proviso that forbade sale or rental to Blacks. Incidentally, such restrictive covenants were made illegal by the federal Fair Housing Act of 1968.

These discriminatory practices had significant practical consequences. Many are unaware that the policies of the government of the United States actively prevented Blacks from securing mortgages for houses in most neighborhoods, not just in Levittown. As mentioned previously, most home purchases involve securing a mortgage from a bank or another suitable financial institution. Prospective buyers shop for a house in the best neighborhood they can afford and buy it, gaining the benefits and suffering the disadvantages of the house and the neighborhood. Of course, the prudent shopper seeks to maximize the former and minimize the latter. The buyer has the best chances of doing so when the range of buying choices is as wide as possible and purchasing power is high. That is why wealthy people live in nicer houses in better neighborhoods than poorer people.

However, the denial of credit to Blacks for the purchase of houses in desirable places, like Levittown, confined them to narrower choices in less desirable neighborhoods. When the time came to make a house repair or improvement, a Levittown resident could go to a bank or other lending institution and get a loan to cover the cost of repairs that would maintain the value of the house or make improvements that would cause its value to appreciate. This was not so easy for a Black homeowner with a house in a Black neighborhood or even a racially integrated one. It was the official policy of the FHA not to underwrite such loans and to discourage financial institutions from doing so. These bars to credit for home maintenance and improvement doomed Black neighborhoods and racially mixed ones to declining careers of disrepair and deterioration. Impelled by the FHA's restrictions on home loans for purchase or repair in Black or racially integrated neighborhoods, financial institutions created maps of residential housing areas all across the United States demarcating, often by circumscribing them with red lines (a policy that came to be known as "redlining"), areas where Black residency made properties within these areas ineligible for FHA loans. Other lenders avoided making loans in redlined areas, too. This made them hard to sell and a struggle to maintain, thereby lowering their value even further.

The policy had particularly pernicious consequences. If residents in all-White neighborhoods sold properties to Blacks, they put the value of all neighboring properties at risk of ineligibility for mortgage or home improvement loans. This ineligibility would lower property values and make it harder to maintain the other houses within the neighborhood. The results were various and all unpleasant. In general, it made many Whites unwilling to sell to Blacks because it would, in fact, tend to lower property values and certainly incur the ire of their neighbors. This kept the pool of houses available for purchase by Blacks small and mostly constrained to already Black neighborhoods. When the first Blacks did manage to buy a house in a White neighborhood, it was likely to be at a greatly inflated price because such buying opportunities were scarce and demand was high. Some in the real estate industry figured out how to play this situation to their maximum advantage. The game, a practice called "blockbusting," went something like this:

Persuade a homeowner in a White neighborhood to sell their house, usually at a significantly higher than market price, to a Black family. Then spread the word that "Blacks are moving in—get out before property values fall." In his book *The Color of the Law*, Richard Rothstein describes the tactics of "creative" real estate hucksters who arranged for Black people to stroll through the targeted neighborhood showing obvious interest in properties, some perhaps pushing baby carriages. He tells how brokers even hired people with speech patterns that many would attribute to Blacks to make calls to houses in the area asking to speak to people with names that "sounded Black." (At that time they would have used names like Leroy or Lacerene rather than Jamal or Jamila.) They hoped these shenanigans would push some homeowners into panic selling, allowing the brokers to swoop in and buy the houses at fire-sale prices, well below the true value of the house. These predatory agents would then turn around and sell the properties at inflated prices to Black buyers eager to move to a better house in a better neighborhood. Almost always, the arrival of several more Black families triggered an exodus among the remaining White residents. Destined for red-lining, the properties were a financial struggle to maintain and difficult to sell. They were more likely to fall in value than to experience even modest appreciation.

An overall result was that Whites who bought into Levittown and others like it secured good housing that was competitively priced and, with the passage of time, appreciated. They made an expenditure for housing that produced sustained increases to their net worth. In contrast, those Blacks who could purchase nice houses often paid inflated prices and found them more of a drain than a significant contributor to their net worth. Since the family home is the biggest part of the net worth of the majority of American families, yesteryear's unequal access to this foundational asset is an important factor contributing to today's inequality in the net worth of Blacks and Whites. Furthermore, the realization that a major factor in generating this inequality was for a long time the de jure discriminatory race-based lending policy of the federal government as well as similar informal efforts by large segments of the private lending sector provides helpful perspective as we take a careful look at why we see such a wide gap in the net worth of Black and White populations.

There is a widespread tendency to attribute much of the difference in net worth between Blacks and Whites to the individual characteristics and behaviors of members of these groups. Many suspect that the lower net worth of Blacks is due to factors such as:

lack of a "strong work ethic," for which the remedy is to work longer and harder

poor spending choices, for which the remedy is to buy houses instead of fancy cars

lack of thrift, for which the remedy might be to invest in certificates of deposit instead of expensive sneakers

In cases where these apply, which are far fewer than the informal advisors offering this gratuitous advice assume, such simple-minded financial counseling might be helpful. However, these problems of intergroup economic inequality are unlikely to be solved by a few "commonsense" remedies and pious nostrums. Our discussion of historical differences in access to homeownership shows one of the many ways in which the explanations are complicated and structural. Much of the ten-fold difference in Black vs. White net worth is largely attributable to five factors: household income (in

2016: $61,200 for White households; $35,400 for Black households); factors related to homeownership; retirement savings; differences in education; and the fact that more Black households are likely to be headed by one person than are White households. All of these factors are important, and the redress necessary to promote greater equality will surely require structural change and the involvement of governmental and nongovernmental institutions.

MONETARY REPARATIONS FOR SLAVERY IS AN IDEA WHOSE TIME WON'T COME

Some have suggested a different, more radical and, they think, rapid approach to leveling the playing field on which Whites and Blacks play the economic game of life in these United States. Every year from 1989 until 2017, his last year in Congress, Representative John Conyers (Democrat, Michigan) introduced H.R. 40, a bill whose summary reads:

> To address the fundamental injustice, cruelty, brutality, and inhumanity of slavery in the United States and the 13 American colonies between 1619 and 1865 and to establish a commission to study and consider a national apology and proposal for reparations for the institution of slavery, its subsequent de jure and de facto racial and economic discrimination against African-Americans, and the impact of these forces on living African-Americans, to make recommendations to the Congress on appropriate remedies, and for other purposes.

H.R. 40 is intended to initiate a program of reparations to Black Americans to partially compensate them for more than 200 years of slavery, during which they worked without pay. It does not attempt to extract punitive damages for the atrocities of physical and sexual violence that were part of the day-to-day life of slave communities, just the wages due for almost two and a half centuries of work as reparations. Some argue that there is a precedent for such a step since the United States has paid reparations to another minority. The story of to whom and why follows.

On February 19, 1942, two months after the Empire of Japan's de facto declaration of war on the United States with an attack on Pearl Harbor, President Franklin Delano Roosevelt issued Executive Order 9066. This decree required removal, by force if necessary, of all persons of Japanese ancestry from California and other parts of the Pacific region of the United States. More than 117,000 persons were relocated to inland areas away from the Pacific Coast and held in internment camps until World War II drew near a close. On short notice and forbidden to bring more than a suitcase or two of personal belongings, Japanese Americans had to sell their homes and possessions and, if they had them, liquidate their businesses.

In 2017, the Franklin Delano Roosevelt Museum located on the Roosevelt estate in Hyde Park, New York, mounted an exhibition titled: "Images of Internment: The Incarceration of Japanese Americans during World War II." This moving and candid collection contains pictures from family albums of some of those relocated and also includes photographs taken by such leading photographers of the era as Dorothea Lange and Ansel Adams. The exhibition provided frank and revealing documentation of what it was like to have your property and business confiscated, your ties to a community and a place severed, and then be forced to move to desolate and remote locations. Photos show row upon row of shabby barren barracks and capture the camp life that included residence in these poorly insulated structures, each shared by many families. There appear to be waiting lines for everything, even to use the communal toilets. There is a picture of a returning Japanese American soldier showing a family member the Purple Heart he was awarded in recognition of the blood he shed for America. Another shows a camp-imprisoned family sitting in front of an American flag, the mother proudly holding a picture of a son in uniform.

In this regard, the exhibit also calls attention to the courage and the military effectiveness of the 442nd Regiment, a unit made up almost exclusively of Japanese Americans that included recruits from the internment camps. The 442nd was the most decorated unit in the American Army. Over the course of its three years of war duty, a total of 13,000 Japanese Americans served in a unit that collected more than 18,000 commendations, including more than 9,000 Purple Hearts and over 4,000 Medals of Honor. Although America lost faith in them because an accident of birth gave them Japanese ancestry, many Japanese Americans did not lose faith in America and took heroic steps to demonstrate it.

Thirty years after the end of World War II the United States agreed to pay reparations to Japanese Americans who were interned and issued a formal apology for its wartime treatment of Japanese Americans. In addition, it was decreed that the ten camps where Japanese Americans were interned be maintained as historical landmarks, reminders of the nation's failure to protect a group of Americans from prejudice, greed, and political expediency. Two kinds of reparation were paid. Every survivor alive when the Civil Liberties Act of 1988 that decreed reparations was passed received a $20,000 payment. However, heirs of those who died before the passage of the Civil Liberties Act received no payment. In recognition of the losses incurred by those who had to sell their property under circumstances of distress with little opportunity to negotiate a fair price, a fund of $1.6 billion, ultimately distributed to more than 82,000 individuals, was allocated to compensate survivors. Although both types of monetary payment were tangible representations of guilt and regret, these sums or even much larger ones could not compensate for the lives interrupted and the psychological damage that was done by

President Roosevelt's relocations. Finally, note the following two features of this program. First, the payments began during the 1980s and were completed by 1991 and were made only to those who had suffered direct harm. Second, many Americans living during the period when reparations were awarded and paid were participants in the civic life of the government that committed the transgressions against our Japanese American population.

Like the proposals of the Japanese American Citizens League that eventually led to the reparation payments mandated by the Civil Liberties Act of 1988 described previously, former Congressman Conyers' proposal suggests reparations for another set of transgressions. There is no doubt that slavery extracted labor without compensation and that it was deliberately destructive of society and family among slaves. Furthermore, enshrined as it was in the Constitution, there is no controversy about the complicity of the United States in the institution of slavery and its maintenance. Some might conclude that reparations to Japanese Americans set a precedent obviously supportive to a case for reparations for slavery. There is no doubt that harm was done to an innocent population who were compelled by another to bear the burden of slavery only because of their African ancestry. Although simply stated in H.R. 40, as we shall see, an exploration of reparations for slavery leads into a complex landscape.

We avoid recommendations for direct monetary reparations as a long-delayed payment for the large and continuing debt America incurred during slavery. There was an opportunity after emancipation when the allocation of "40 acres and a mule" was considered as a grant to aid the transition of former slaves from dependency to self-reliant independence. As the Civil War drew to a close, some recognized that mere freedom without the economic means to sustain life was incomplete emancipation. A grant to freedmen of forty acres and a mule was discussed but never written into law as an entitlement. It was too bad. This was a realistic and practical proposal that might have proven a workable, if niggardly, acknowledgment of the debt incurred by slavery. At that time, the perpetrators and the direct victims of the crime of slavery were alive and could be identified. Today, both have been gone for generations.

The United States now includes citizens of color whose ancestors had not yet settled in the United States—think of East Asians, Americans of Middle Eastern extraction, or populations of the Indian Subcontinent. Would they feel a sense of obligation? Would they feel any sense of collective guilt? Probably not. For that matter, there are many among today's White population who understand that slavery was a crime but feel no complicity in its commission. Consequently, they would not feel it either appropriate or just for the government to take their money and give it to Black Americans who never suffered slavery and live in a country where it has been illegal to own people for more than 150 years. While it is true that, even today, White

Americans and many other Americans, too, benefit from fruits of labor stolen from Black slaves, that argument opens doors to many complications.

Many Black Americans live on lands that were taken from the First Americans, the native populations. How would those Black Americans view reverting these properties to the descendants of native populations? Many of us are familiar with the phrase "trail of tears" used to denote the forced migration of Native American tribes such as the Cherokee from the Southeastern United States to western territories such as the territory that would become Oklahoma. Monetary compensation for the real estate the Cherokee and other tribes left behind in North Carolina, Georgia, and Florida would be a substantial sum. However, because some of the Cherokee were holders of African slaves, whom they took West with them, how much of a reduction in the amount of land they should return to Native Americans would today's Black Americans claim? But on the other hand, after the Civil War, some African Americans voluntarily fought in the campaigns to take more Indian land and to maintain repression of Native Americans in the West. These regiments of Black cavalry troops were known as "buffalo soldiers." They were highly effective, and some of their members fought with sufficient tenacity and ferocity to win Medals of Honor, one of the army's highest decorations and awarded only for distinguished service. In settling reparations accounts, would Black participation in these acts of Indian oppression reduce the debt Indians owe Blacks for participating in their enslavement? Would the oppressive and lethal deeds of the buffalo soldiers perhaps even increase the amount of land some Blacks should surrender back to some Native populations?

Otherwise a silly waste of time, these absurdities are useful to illustrate how tangled and difficult it is to appropriately and fairly levy reparations on current generations to pay for the transgressions of generations long gone. How does one decide who should get a check or who receives a deed? Would President Barack Obama get a check? Would the wealthy Mashpee Wampanoag operators of a lucrative casino business get more land, perhaps using it to build still more casinos? Many East Asian and South Indian Americans might object to the use of their taxes to support reparations to either Blacks or Native Americans, justly claiming that their populations had no role in grabbing Native American lands or in the enslavement of Blacks. While the outrages inflicted on Blacks and Native Americans are manifest and beyond dispute, it would be quite challenging to justly levy the costs of reparations and extraordinarily challenging to distribute reparation checks or parcels of land equitably. Reparations as redress of depredations now interred in history are certain to be extremely difficult. Such a policy is equally certain to further divide us into antagonistic tribes—creating conflict instead of resolving it. Instead of being accepted as just redress, such efforts would be widely viewed as illegitimate, generating acrimony and inspiring determined, perhaps nasty, resistance.

NOTES ON THE BIRTH, CAREER, AND LIKELY DEMISE OF AFFIRMATIVE ACTION

Instead of reparations, a better case has been made for some forms of affirmative action. This innovative social remedy was originally intended to prevent and redress discrimination by targeted extension of opportunities to groups whose progress has been historically compromised by discrimination. Here in the United States, Blacks were the initial target of affirmative action. Over the years since its inception, it has been broadened to include Hispanics and some other groups, most notably women and, finally, those with disabilities. Prior to affirmative action, Blacks were largely or completely excluded from many employment sectors and institutions. Not too long ago, a look around the campuses of Princeton, Harvard, or Yale would have found only a few Black students and no Black professors. In the private sector, there were no Black executives at IBM or Bank of America; in government, there were no cabinet members; in major philanthropic organizations, there were no Blacks in the executive suites of the Ford or Rockefeller Foundations; and they were absent from mainstream print and broadcast media, too. There would have been no women, either.

Things are very different now. Breaching the dam of discrimination, affirmative action has released a torrent of talent that has flowed into government, business, manufacturing, the armed services, universities, and numerous other organizations. Certainly, many Blacks, Hispanics, and Native Americans who would have been outside the mainstream are now part of the flow. Women, especially White women, are the group that realized the largest gains from affirmative action. Clearly, affirmative action has affected the pool of talent. It's bigger and broader, at last including whole categories previously excluded.

Affirmative action, a gift of the civil rights movement, also changed hiring practices. It is now expected that positions will be publically and widely announced. Affirmative action was originally intended as a counter to a once pervasive exclusion of Blacks from applicant pools for jobs in the private sector and in government, as well. The result was that Blacks were not only refused employment, but they were also invisible and not considered. As realization dawned that Blacks were not the only invisible group, affirmative action evolved to address inclusion more broadly, seeking to assure that applicant pools included individuals from many segments of American society; yes, this included Blacks but also Hispanics, Asians, and certainly women, too. The "affirmative" in affirmative action carried an expectation that an effort would be made to encourage a diversity of applicants to apply. Today, it is routine for employers to at least pay lip service to generating a diverse applicant pool. Yesterday, it was not.

Granting its many and important positive contributions to society, objections have been raised to affirmative action. Affirmative action means more than announcing and opening an opportunity to those who in earlier times would have been denied. Affirmative action programs that fail to produce more hires of minorities and women or that do not enroll more students from targeted populations are considered unsuccessful. This is intended to put pressure on businesses, institutions, and governments to increase their hires of individuals belonging to these groups. If membership in a group is a significant factor in being chosen for a benefit, all other things being equal, lack of membership could be a disadvantage. In competitive situations where the benefit is limited—who gets the job or who gets admitted to this college or that medical school—a new layer of concern about fairness is introduced.

However, let's be clear about fairness issues. Historically, various types of old-boy networks determined who heard about the jobs and who got the jobs. Whether we are talking about a position in a white-shoe law firm, a job as a fireman, or a place in the freshman class at Harvard, it was an old (White) boy network that excluded women, Asians, Hispanics, Native Americans, and certainly Blacks. Aside from being White and male, qualifications for inclusion in a particular network varied. Money, social position, and a relative who was an alumnus were the usual qualifiers for inclusion in the networks that were helpful in gaining admission to an elite educational institution such as Harvard. One of us once asked Elting Morison why he went to Harvard during the Depression years. He answered, "Because my father had a thousand dollars and Morisons have always gone to Harvard." This conversation took place long ago when Morison and some of his faculty colleagues at Yale were part of a determined and largely successful effort to democratize the Yale experience. His remark was intended to highlight the inequities imposed by the birth-bestowed privileges enjoyed by some.

Not too long ago, if one moved from the ivied walls of Harvard to the firehouses of Cambridge, in the days before affirmative action, new firemen were usually family or friends of old firemen. Old and new firemen were White. Few would be shocked to learn that high-paying blue collar jobs such as electrician were controlled by a union, in this case, the International Brotherhood of Electrical Workers, whose locals functioned as old-boy networks that restricted or did not admit Blacks. These examples could be multiplied, and there were many areas of employment where for many years the restrictive practices or policies of unions or other gatekeepers excluded Blacks. In the next few paragraphs, as we think about the potential of affirmative action to generate inequities, it is important to be aware that the level of unfairness traceable to affirmative action is paltry compared to that widely imposed for so many years by racial, ethnic, and gender barriers.

Nevertheless, when affirmative action is believed to choose one person over another without merit being the determining factor, the practice is perceived as unfair or even illegitimate. And indeed, in some instances, it should be so regarded. However, important factors larger than the simple comparison of the resume of this individual with the resume of that one may be involved and justify the conviction that affirmative action resulted in choosing *the best person for the job*.

To clarify, suppose a distinguished, old law firm in Manhattan says for years that it welcomes women. However, none have been hired. Perhaps the actual hire of a well-qualified woman is the only way to convince others, and perhaps themselves, that the door is really open. Alternatively, consider a large police force that polices a Mexican American neighborhood but that has never had any officers who are Mexican American on the force. The hire of some qualified Mexican Americans would demonstrate a willingness to include members of the community it serves. In addition, the police force gains a greater understanding of that neighborhood. In both these situations, persons fully qualified to do the job may be selected from a pool containing people with better credentials because they are better able to help the organization achieve a legitimate and desirable social or organizational goal. In short, they are chosen because they are *the best for the job*.

Affirmative action in college admissions has fueled dispute for many years. There is a perception, which is widely held in some circles, that Black students with inferior credentials are being admitted with the result being that some White and Asian students are being denied the admission they deserve. This appears to be a big argument over small stakes, at least with regard to Black enrollment.

In the most selective private universities and colleges as well as in flagship state universities, Black students are a small fraction of the students admitted, and their percentage most often is in the single digits. A 2017 survey published by the *New York Times* showed that Black students comprised 9 percent of freshman enrollment at Ivy League universities, 3 percent of freshmen in the University of California system, and 6 percent of freshmen at flagship state universities. The nation's top liberal arts colleges enroll freshman classes that are about 7 percent Black. In general, Hispanic enrollments are a little more than double the enrollments of Blacks, except in the University of California system where Hispanic enrollment was around ten times higher than that of Blacks. In 2017, Hispanics were 39 percent and Blacks were about 5.6 percent of the population of California. These numbers demonstrate that worries about large numbers of deserving Whites and Asians being denied admission to these and other colleges and universities due to race-based admission policies are unfounded.

However, a race does not receive a letter of acceptance or denial from a college or university admissions committee. These letters show up in the mailboxes of individuals. Whatever their race, individuals who have worked hard and capably to earn admission to a college of their choice and are not accepted will feel hurt and disappointment. The perception that someone less deserving gains admission rubs salt in the wound inflicted by denial and evokes resentment. Although there are many targets against which disappointed applicants might direct their ire, race-influenced admissions have emerged as the target of choice. However, if those aggrieved applicants had access to the discussions and records of the admissions committee, they would be likely to find that some White or even Asian students they would consider less deserving than themselves were also admitted. After all, in addition to grades and test scores, college admission is influenced by other factors, such as actual or potential financial donation to the institution, family legacy, ability to play a particular sport, and, sometimes, gender or race. During the spring of 2019, headlines and newscasts carried accounts of fraudulent admissions practices that mostly benefitted upper-middle-class and wealthy White applicants. It is clear that disappointed applicants have a target-rich environment. Nevertheless, lawyers and their clients usually choose to shoot at race, often aiming at a "Black" bull's-eye.

Perhaps this is understandable. Some things can be changed, but others cannot. It is possible, at least for some families, to contemplate raising levels of legal contribution to an institution, family legacy status might be modified by strategic marriage, and there can be a choice of what sport to play. Since it is possible to change gender, a male desperate and determined to enter Smith College could choose to undergo gender-changing surgery, presumably emerging as an admission-eligible female applicant. Race is different. Back in chapter 2, we offered examples that demonstrated the difficulty of legitimately changing one's race, and we saw the price paid by those who attempted to do so and were found out. Of all the nonacademic factors influencing admission decisions, race is one that can't be changed. Consequently, those who experience race-based inequalities feel wronged because of an accident of birth.

However, the landscape is complex. Colleges and universities set out to do more than teach engineering or English literature. They intend to inculcate and reward habits of inquiry, industry, and integrity. They create an environment where those enrolled learn something about the lives and personalities of other students and of those who instruct them, too. In turn, they themselves are living texts, studied, learned, and better understood by others. When the students and faculty reflect variety in the larger society world beyond the campus, there is an opportunity to learn more broadly. In such diverse social settings, experience can show how fundamentally similar people are and encourage awareness and appreciation of the real, important, and often valuable differences that exist between ethnic, national, and racial groups. Colleges and universities acknowledge this obligation to educate beyond subject matter and discharge it by intentionally creating diverse student bodies and struggling to build diverse faculties to teach them. The greatest progress toward achieving such carefully crafted diversity has been achieved on the campuses of America's elite private institutions. It has to be acknowledged that sincere pursuit of such diversity requires the selection of students and the recruitment of faculty with an eye that is not entirely blind to culture, color, or gender. Slots in the student body or on the faculty of these outstanding institutions are highly prized. The criteria for selection to the freshman class or to the faculty are strongly influenced by merit.

However, as mentioned earlier, in some circumstances and for some purposes, selection committees may consider both who appears to be the best person and who seems to be the best person for the slot. In most cases, the same individual matches both criteria. There are cases where they do not. As an example, consider a mathematics department that teaches legions of undergraduates and scores of graduate students but has never had a female faculty member. When they advertise a faculty position, the applicant pool generated contains many well-qualified male applicants and some well-qualified female ones. The resumes of the strong male and the strong female applicants indicate that all are likely to handle responsibilities of teaching and research capably. Presented with such a situation, the department may choose a strong female applicant over a stronger male one. In extending a job offer to the woman, the institution may have chosen *the best person for the job*. By hiring a well-qualified female applicant, the department was able to provide students the experience of learning mathematics from a highly capable woman. Their female students, some of whom may be contemplating a career in mathematics, experience a concrete demonstration that such a career path is realistic and open to them. Male colleagues in the mathematics department gain an opportunity to see that female colleagues can work out just fine. On the other hand, if search committees are willing to use affirmative action just to "check a box" and, forsaking quality, hire only with regard to gender or race, there is the likelihood that their new hire may teach the wrong lessons. Some students may come away from her classes with the false generalization that women are less capable at mathematics than men. Some faculty in the mathematics department may also have gender biases reinforced. However, the use of affirmative action to bring the right women to the mathematics department can educate at many levels and in many valuable ways. Nevertheless, while valid justifications for affirmative action can be offered, in this case, an otherwise worthy candidate may have been passed over because his Y chromosome prevents him from satisfying some broader institutional priorities.

Under intensifying siege, affirmative action may be approaching the end of its game-changing, if controversial, career. It has been an important contributor to the change in the complexion and gender of many employment sectors and in higher education. It has, for example, increased the number of people of color and women involved in public safety sectors. Despite the irony that White women have been its major beneficiaries, it has been a key force in unlocking the doors of corporate suites and increasing the flow of well-qualified people of color emerging from the nation's top colleges and universities from a feeble trickle to an impressive stream. How much longer affirmative action will be a major tributary of that stream is very much in question. Initiated by the Kennedy and Johnson Administrations to include, it now increasingly divides. The fall of 2018 saw people of color, an association of Asian Americans, attack what they perceived as unmerited admission to an elite university of other people of color, Hispanics and Blacks.

Michael Wang became the engaging poster boy of an Asian American challenge to affirmative action. Michael was an unusually strong student, with a very high grade point average and high test scores, he also sings and plays piano and was a member of his high school choir when it sang at Obama's inauguration and also when it performed at the San Francisco Opera. He was a school leader and excelled at debate and public speaking. A *New Yorker* article by Hua Hsu, a staff writer at the magazine, reports that he was a cofounder of his high school's math club. In light of all this, his expectations of admission to some of the nation's most competitive universities seemed quite realistic, almost inevitable. He applied to Yale, Princeton, and Stanford and was rejected by all three. His disappointment turned to anger when he learned that some of his Hispanic and Black classmates whose credentials he perceived as inferior to his own gained admission to one or another of the elite trio of schools just mentioned.

This prompted Wang to write an op-ed to the *San Jose Mercury*, a widely read and influential newspaper in the Bay Area where he lived. In this piece, he cited studies indicating that Asian American applicants to elite institutions needed scores on SAT tests 140 points higher than Whites, 270 points higher than Latinos, and 310 points higher than Blacks. He concluded that race-conscious admission policies practiced by these institutions worked to the disadvantage of Asian Americans and urged adoption of race-blind admission policies.

Objectively viewed, since so many more Whites than any other group are admitted to these schools, one is tempted to ask: Wouldn't Asian Americans do even better by targeting the White applicant pool rather than focusing on the minority ones identified in Michael Wang's editorial? After all, so many more Whites are accepted, and their scores are also significantly lower. At any rate, as much as any document, Wang's op-ed piece was an inspirational and key forerunner of a movement called Students For Fair Admissions

(SFFA). Four years later, the fall of 2018 saw SFFA accusations that Harvard's failure to use race-neutral admissions policies is a major contributor to discrimination against Asian applicants having their day in federal court.

The passing of affirmative action need not mean the death of progressive measures to correct widespread disparities in access to and participation in higher education. In the twenty-first century, race-based allocation of benefits such as college admission is challenged by the increasing disparities of wealth, education, and practical power within a Black population that is not homogeneous. One of the many insights found in *Disintegration*, Eugene Robinson's informative and Pulitzer Prize–winning book, is that there are growing disparities of wealth and education within the group racially identified as Black. Increasingly, middle- and upper-income Blacks are the beneficiaries of programs their creators had hoped would strongly impact all Blacks. This is not due to some sort of corrupt fix. Among Blacks, as in other groups, those in more advantaged circumstances are more likely to be aware of new opportunities and to have the capacity to mobilize the social and economic resources to exploit them than those less advantaged. While there is little doubt that affirmative action has been an important factor in increasing the strength and the size of the Black middle and upper classes, the good news is that these subpopulations are increasingly quite capable of holding their own.

At this stage, much more will be accomplished by focusing resources on the truly disadvantaged. Rather than being directed at particular racial or ethnic categories, policies to improve educational opportunity and access should address individuals, whatever their racial, gender, or ethnic affiliations, according to need. Metaphorically, we should fix the potholes wherever they are, not just on Elm Street and other streets named after trees. Beyond fairness, broadly applied and inclusive of all in need, such strategies are politically wise and more likely to gain and to deserve wide acceptance.

BLACK LIVES AND OTHER LIVES ENDANGERED AND COMPROMISED BY LAW ENFORCEMENT AND ITS AGENTS

While opinions may differ on whether monetary reparations or affirmative action are appropriate ways to repair some of the injury inflicted by slavery and discrimination, all should agree that equality under the law and equal treatment by agents of law enforcement are essential parts of the solution to our race problem. Before the civil rights acts, legally mandated segregation and abridgment of voting rights were the systemic indicators of our failure to extend equality under the law to all our citizens. Landmark court rulings and national legislation abolishing de jure school segregation, granting access to public accommodations and enforcing voting rights were long and indispensable steps toward equality under the law.

However, the justice system still collides with Blacks, Native Americans, and some Hispanic populations in ways that are destructive, often unfair and brutal, sometimes even fatal. Some police do not enforce the law equally, the penal system incarcerates too many for too long, and prosecutors can be selectively zealous and have, on too many occasions, been far from color blind. These disparities have affected minority communities in a variety of ways. People who are mistrustful or fearful of police and other arms of law enforcement are less likely to provide information that might be helpful to police in preventing crime or apprehending perpetrators. This is too bad because their tips could aid in the identification and arrest of perpetrators of crimes that make their community and other communities less safe. In addition, their testimony could aid in the prosecution of offenders.

For some communities, broad patterns of incarceration have made a stint in prison almost as unremarkable as attending high school and more common than military service. Although the corrosive impact of selective or discriminatory law enforcement on the social structures and the economic well-being of impacted communities is well-established, the benefits to the public safety of those communities are far from comparable to the disruption imposed by the mass incarceration that Michelle Alexander has highlighted in her eye-opening and revelatory book, *The New Jim Crow*. At a larger societal level, the costs of our prisons have produced a low return on the considerable investments we continue to make. Reflection on the difficult and tragic examples explored below make it hard to ignore the claims of racial disparities in our system of policing, enforcement, and incarceration. The injustice and ugliness of the problem is illustrated by the stories that follow.

Around 9 p.m. as Philando Castile, his girlfriend, Diamond Reynolds, and her four-year-old daughter, Dae'Anne, were driving along Larpenteur Avenue in Falcon Heights, a suburb of St. Paul, Minnesota, the flashing lights of a patrol car signaled them to pull over. This was not an unusual experience for Castile. Over the past several years he had been pulled over more than fifty times and accused of violations, many of them minor or technical. The stopped vehicle was approached by two police officers, one on each side of the car. The officer on the driver's side directed Castile to produce his license and registration. Castile told the officer that he was wearing a firearm. Castile had a permit to carry and was legally doing so. At that point, the officer directed that he should not pull it. The motorist said he would not. The officer unholstered his weapon, again said, "Don't pull it out," and then fired his service revolver into the car seven times. Reynolds, who was seated next to him in the front passenger seat, and her young child sitting on the back seat both escaped injury, but Castile was hit by five of the shots, two of which struck his heart. A little over forty seconds elapsed between the time the officer first spoke to the motorist and the time he fired several shots into a car with a man, a woman, and a small child.

The officer said he was in fear for his life. However, his partner, who was at the passenger-side window, did not draw his weapon or behave in a manner that would have suggested the situation was one of great danger. After the shooting, neither policeman attempted to aid the victim. The officers who arrived at the scene shortly after the killing focused their attention on comforting and reassuring the shooter, a fellow officer, ignoring the victim who died later in a hospital emergency room. Although making no effort to stabilize the victim, they forced the woman out of the car, ordered her to her knees, handcuffed her, and took her into overnight custody. Within a day or so, as the circumstances surrounding this shooting came to light, numerous public officials, including the governor of Minnesota and a US senator, forcefully spoke out about the injustice done to Castile. The officer was tried for manslaughter and reckless discharge of a firearm, but a jury did not convict him. However, subsequently the Castile family was awarded a settlement of $2,995,000 and his girlfriend, Ms. Reynolds, received a settlement of $800,000 providing both symbolic and significant tangible recognition of the officer's wrongdoing.

Keeping this story in mind, contrast it with another shooting story. This one unfolded a little after midnight at a Century 16 multiplex in Aurora, Colorado, when James Holmes, a graduate student in psychology who had recently dropped out of the University of Colorado, left the theater and went to his car in the parking lot. He returned wearing body armor and carrying gas grenades, a 12-gauge shotgun, a handgun, and an automatic rifle. He interrupted the midnight screening of *The Dark Knight Rises* by releasing tear gas grenades. He then unleashed a barrage of shotgun and rifle fire on the audience. When he finished his rampage, the toll of this horrific mass shooting was twelve killed and fifty-eight with gunshot wounds. After leaving the theater, he was apprehended by police in the parking lot. He did not resist arrest and was handcuffed and taken into custody. Despite the fact that the police knew they had rushed to what was probably the scene of a mass shooting, they acted with admirable discipline and arrested the shooter without firing a shot or resorting to assaults such as chokeholds or other means of violent restraint.

These crimes differ greatly in the magnitude of the crimes the police perceived the perpetrator to have committed. In Minnesota, it was a traffic violation and perhaps a vague suspicion the driver might have a slight resemblance to the perpetrator of a robbery that took place on another occasion. In Colorado, police knew they were arriving at the scene of a mass shooting by a heavily armed, clearly dangerous, and possibly deranged man. They also differ in the race of the perpetrator. Holmes, the theater killer, was armed to the teeth and had just shot up a theater. He is White. Castile was a thirty-two-year-old Black man who was driving a car with a broken taillight. He might have borne a slight resemblance to a robbery suspect and was legally carrying a handgun he announced but did not unholster or brandish. In 2012, the White mass killer was taken alive, arrested, and, in 2017, sentenced to life without parole. Castile, guilty of a misdemeanor at worst, was fatally shot five times within less than a minute of being stopped and was buried a week later.

The tragedy of Philandro Castile is just one of the far too many examples of the excessive and lethal force police deploy against Blacks. Many found it difficult to believe that Tamir Rice, a twelve-year-old, was shot dead in a park by police while brandishing a toy gun. However, he was, and they buried him, too. In Charleston, South Carolina, Walter Scott was fatally shot in the back by police while fleeing the scene of a traffic violation. In the New York City borough of Staten Island, Eric Garner was caught illegally selling cigarettes on the street. He was assaulted by a posse of overzealous police officers and, in front of onlookers, choked to death—"I can't breathe" were among his last words.

There are so many stories, too many to recite here, of Blacks who have been victims of excessive, and too often lethal, police force. Most are disturbingly similar and fall into a pattern that features a Black committing, or being suspected of committing, an infraction that is not an immediate threat to the life of police or civilians. In many cases the infraction is minor. Subsequently, a confrontation with law enforcement results in a one-sided escalation of force by police that quickly ends in the death of the Black suspect. Police usually said they feared for their lives or that they viewed the situation as one of "kill or be killed." In one of those cases where the suspect turned tail and ran, the officer apparently decided that the escape of a fleeing suspect posed such a danger to society that lethal force to prevent escape was appropriate.

Headlines and the eleven o'clock news put the focus on police killings of Blacks. Many will be as surprised as we were to learn that Native Americans are at even greater risk of dying at the hands of police than Blacks. The numbers show that in 2016 Native Americans were killed at a rate of 10.13 per million; Blacks at 6.66 per million; Hispanics at 3.23 per million; Whites at 2.9 per million; and Asian/Pacific Islanders at 1.17 per million. Just as police killings of Blacks inspired the Black Lives Matter movement, the disproportionate killing of Native Americans by police has generated a Native Lives Matter movement, too. These loosely organized but increasingly influential associations are forcing attention to correcting the disparate treatment of these populations by law enforcement.

As awful as the eager, almost wanton, use of lethal force by police is for disadvantaged communities of color, as Alexander has explained in *The New Jim Crow*, incarceration has had even more pernicious effects. Compared

with those experienced by the White population, incarceration rates of citizens of color are shamefully high. At the time of the 2010 census, Whites
were imprisoned at the rate of about 430 per 100,000, while rates for Blacks,
Hispanics, and Native Americans were 2,306 for Blacks, 831 for Hispanics,
and 895 for Native Americans per 100,000. In all groups, incarceration rates
of males are much higher than those of females. A little over half the
American population is female, but the prison population is only about 7
percent female. While in aggregate these numbers are large and depressing,
we need to consider their social implications.

Incarceration affects the inmate, his family, and his community. While in
prison, the inmate loses income, and whatever he contributed to the support
of others, that is lost, too. Family and social relationships suffer. Interrupted
by incarceration, they are often loosened, frequently dissolved, and almost
never strengthened. When they are released back into society, those who
have been imprisoned find it difficult to compete with those who have not
been for desirable jobs that offer a good starting wage and possibilities for
wage growth. When they reenter society and finally find work, the wages
they receive are likely to be lower and they will have lost their voting rights.
The jobs they find are also likely to be less secure. The combination of the
prison experience, hardly a promising one for the development of wholesome
social networks, and the difficulties of finding work after release increase the
likelihood that someone seeing himself without other choices will turn to
income-producing criminal activities. The weight of so many men in some
Black communities limited by backgrounds of incarceration is a social burden that the members of those communities, especially their families, must
struggle to carry. However, perhaps the most important problem is poverty,
the soil from which high incarceration rates and other problems faced by
these communities grow. High rates of incarceration compound the problems
of communities where incomes are too low and people just don't have
enough money.

MORE EDUCATION AND TRAINING IS
THE SUREST PATH TO HIGHER INCOMES

In thinking about this problem, perhaps it would be useful to return for a
moment to the American Jewish population, a group we suggested might be
the most broadly successful population in the United States and is among the
highest-income groups in the country. Those who know the history of the
Jewish population in America are aware that this has not always been the
case. While Jews have had a presence in the United States since the colonial
era, at that time, their numbers were small. The arrival of a wave of largely
German immigrants after the middle of the nineteenth century increased the

population from a few thousand to a couple hundred thousand who established stable and thriving communities across the United States, many starting small businesses, some rising to middle-class status and some acquiring college educations.

The largest expansion of the US Jewish population began with the arrival of large numbers of Eastern European Jews, many fleeing repressive and dangerous conditions, during the latter part of the nineteenth and early twentieth century. The population swelled into the millions. However, for the most part, these new arrivals were poor. Since the Jewish communities in which they grew up expected its male members to be able to read the Torah, the men were literate, but most were not well-educated. On arrival, they found living here in the United States hard and challenging; most flowed into the lower echelons of the working class. Within a generation or so, they began to prosper, and by the 1940s, they were well along the way to becoming the highly educated and broadly successful population we described earlier. What happened? What propelled this metamorphosis?

There were many factors, and these included the social and material support of many Jewish community associations, such as the Hebrew Immigrant Aid Society. Immigrant Jewish populations were part of a broadly shared religion reinforced by a variety of organizations. This provided a level of social cohesion that was enjoyed by few other immigrant groups. In addition, there was a broad subscription to programs that aided the learning of skills that would aid in vocational and business activities. However, more than any other, the key factor was education, of many kinds. Free public education was not a feature of the social landscape left behind by Jewish immigrants arriving from Eastern Europe, and the opportunity to educate their children was seized upon.

To a greater degree than other Americans, Jews went to college and completed it. Carmel Chiswick has pointed out that by 1940 there was a clear difference in the educational levels of the American Jewish population and its non-Jewish one. A survey of Jewish men indicated that more than a quarter of those born before 1940 had college degrees. Furthermore, several of these had postgraduate professional degrees in areas such as law and medicine. Strikingly, this was more than twice the percentage in the non-Jewish population and provided a telling indicator of the priority placed on education by this segment of American society. This population's early and continued investment in education is a major reason why more than 66 percent of Jewish men, as opposed to 20 percent of non-Jewish ones, are in higher level occupations that include the professions and management, occupations that award incomes in the upper and top levels of income distribution. Importantly, the good news is that there is nothing Jewish-specific about the power of education to lift groups into the upper percentiles of the income distributions. A look at the data shows it works for many groups.

South Asians and East Asians are groups that also earn high incomes. The delightful and hilarious romantic comedy *Crazy Rich Asians* suggests marrying someone rich provides an avenue to get rich. However, the vast majority of Asians in America have found a much better and more broadly accessible route. It leads through the classroom to graduate and postgraduate degrees. Among Asians, the highest family incomes are earned by Indian families, where it is not unusual for both husband and wife to have advanced degrees. We should note that in the South Asian population, the choice of fields in which advanced degrees were pursued was not randomly distributed but was often in the professions, such as law or medicine, or in the fields of STEM (science, technology, engineering, and mathematics). Justly or unjustly, STEM degrees generally bring higher incomes than those in the arts and humanities. Those in professions such as medicine earn multiples of the average or median income.

Within the Black population as in others, education also makes a big difference in income. In 2009, Black men without a high school education had a median annual income of around $30,000. A high school diploma added another $4,000, raising the figure to $34,000. Acquiring a bachelor's degree boosted the income to $40,100, 117 percent of that earned by the median Black male high school graduate. Obtaining an advanced degree had much greater impact, yielding a $21,400 increase to $61,400, boosting the median income to 153 percent of the median of those who did not advance beyond a bachelor's degree. These are significant differences and wider holding of bachelor's and advanced degrees would make a significant impact on the ability of this population to increase its net worth. Clearly, these numbers make a persuasive case for continued and determined effort to raise the percentage of Blacks completing higher education. In 2015, the US Census Bureau reported that 22.5 percent of the Black population had received bachelor's degrees, and 8.2 percent had earned an advanced degree. These numbers contrast with higher numbers in the White population, of which 32.8 percent hold bachelor's degrees and 12.1 percent have advanced degrees and are much lower than the 53.8 percent of Asians who have bachelor's degrees and the 21.4 percent that have secured advanced degrees.

MORE THAN EDUCATION WILL BE NEEDED TO CLOSE THE ECONOMIC GAP BETWEEN BLACKS AND WHITES

More education will narrow the gap between Blacks and other populations, but it won't bring Black families to economic parity with White families. One of the reasons is quite simple. Family net worth is an aggregate of the wealth of all members of the family household. Since around 60 percent of Black families are headed by one person and about 30 percent of White families are headed by one person, even if Blacks and Whites earned equivalent incomes, and we know they don't, there would still be a mismatch in family net worth. In addition, we know that discriminatory structural factors, some of which we have discussed, are important contributors to Black–White wealth disparities. However, we must not overlook the entanglement of family structure with education and net worth. Within both Black and White populations, marriage rates are lowest among the least well-educated. One thing is often connected to another. In this case, lower levels of education predispose to lower incomes, and as studies have shown, lower levels of income lower marriage rates, too. To the extent that children of single-parent households are more likely to be born into circumstances of low net worth, it becomes less likely that the educations they receive will be of high quality and include college. Such a sequence of events has pernicious consequences since it is likely that the economic trajectories of children who do not receive college degrees or advanced vocational training will be lower.

In addition to less income, single-parent families struggle with many disadvantages two-parent families don't. They don't have the advantages of division of labor—one partner gets the car registered, the other does the laundry; one partner forfeits pay to make a school visit, but the other stays at work; dad takes one child in for a medical procedure, mom takes care of the other two and fixes dinner. Among the ways to reduce Black–White differences in net worth, a descriptor that entrains many things including education and health care, is fostering social conditions that enable and encourage the formation and maintenance of two-parent rather than single-parent families. In 2015, the marriage rate in the Black population was around 30 percent, only a little more than half the 54 percent seen in the White population. It is true that marriage rates have been declining in many industrialized nations. However, to the extent that it results in single-parent households, the impact of this change in social mores on net worth should prompt Black Americans to struggle to maintain two-parent households.

THE BLACK VOTE—PUT IT IN CONTEST

The ballot remains a powerful tool for influencing social policies, and its strategic use by Black voters will help effect progressive policies that will lift Blacks and other groups in their struggle to gain the three Bs: better wages, better education, and better living conditions. The three Bs will facilitate the movement of Blacks, Native Americans, and Hispanics toward greater equality with the White, and still majority, population of the United States. Wisely crafted and equally applied on the basis of need, progressive social policies will lift neglected segments of the White population, too.

Consideration of how to give the Black vote the greatest impact triggers recall of an eye-opening conversation with William Galston, a Senior Fellow at the Brookings Institution and weekly columnist for the *Wall Street Journal*. During our chat, Galston identified some striking parallels between Jewish voters and farmers. Both groups are numerically small and comprise only 2 to 3 percent of the population. However, both are potent political forces in American politics. Neither group is uniformly distributed across the United States but instead concentrated in particular states and regions. Farmers are very strong in the agricultural Midwest and also influential in parts of the South and agricultural regions of some other states, including California. The voting strength of the Jewish population is found in the several states of the East Coast and in some big Midwestern population centers such as Chicago and places such as Los Angeles and the Bay Area in California. Farmers and Jews influence the politics and the political leaders in those areas where they are concentrated, and the politicians representing these areas influence the shape and direction of national policies. Both groups are knowledgeable about the stands of candidates on their core issues.

Moreover, as surely as the sun rises in the east, they vote. In this way, they may reward those who champion their interests and punish politicians whose stances or legislative actions run counter to those interests. While farmers lean Republican, Jews are more likely to support Democrats. However, neither group is firmly in anyone's pocket. This is understood by both parties, and to varying degrees, Democrats and Republicans court the support of these small but politically potent populations.

Since the Roosevelt Administration and its New Deal, Blacks have increasingly voted Democratic. However, both of us recall a time when many Blacks were Republicans. For some, it was a nod to President Lincoln, a Republican who freed the slaves, and until the Johnson Administration, the implacable racism of the Democrat's Southern wing kept many Black votes in the GOP column. Today, the Black vote is overwhelmingly Democratic. In fact, Blacks are the Democrats' most dependably loyal constituency. At least it is when it votes. Perhaps importation of the voting habits and strategies of America's farmers and Jews would be useful in the drive for full Black equality. The two elements Blacks should be most keen to appropriate are independence and voter participation.

Both independence and voter participation are essential and must be deployed in combination. Independence encourages politicians of both parties to really listen to your concerns, and it provides an incentive for them to address, or promise to address, those concerns provided they can be woven into the particular political fabric of their party. Voter participation serves as the indispensable carrot or stick. Only by turning out to vote can a group reward a candidate for furthering their issues, and only by voting for their opponent can the group punish politicians for failing to do so. Returning to independence, it carries an important defensive benefit. It can moderate the degree to which a candidate might act or promise to act against a group's interest. Blacks must be aware that a politician who has no hope of winning any Black votes need not moderate a stand to avoid offending or pushing away voters he or she had no chance of getting anyway.

SUMMING UP

In this chapter we have identified areas of inequality between racial groups, particularly between Blacks and Whites, because we believe extension of equality of opportunity and equal treatment to all will be the foundation on which solutions to America's race problems will stand. There are many "race problems," some of which were discussed in this chapter, and others that have been addressed elsewhere in this book. However, in the United States, the problems rooted in race involving Blacks, Whites, and a still White-dominated governmental apparatus are our most acute. It is this set of durable, difficult, and still unsolved race problems that are most prominently addressed here. We have highlighted economic disparities, educational disparities, unequal treatment by the agents and institutions of law enforcement, and health-care disparities. We know that racism in its various forms is a pervasive and disruptive influence that denies equality, generates inequalities, and asserts false hierarchy. However, taking a cue from Dr. Martin Luther King Jr.'s perceptive observation, "morality cannot be legislated but behavior can be regulated," we do not attempt to write prescriptions that will cure racism. After all, racism is a mental condition, and there are no FDA-approved drugs or devices for its treatment. Instead, in this chapter and others, we have identified some measures and behaviors that will help reduce inequalities and foster equality of treatment and opportunity. Although these measures will not eliminate racism, we believe that taking the steps and instituting the policies we have discussed will treat some of the injuries it has caused and insulate against its corrosive effects in the future.

Narrowing the gap in net worth between Black and White populations will be required for Blacks to gain full equality. In this chapter, the foundational role of economics, discussed as inequalities in net worth, was shown to underlie or strongly influence the quantity and quality of education received by the nation's citizens, to have a major impact on their interactions with the legal system, and to be a determinative factor in the kind and quality of their health care. Important steps toward equality will involve correction of the inequalities Blacks experience at the hands of the agents of law enforcement and in its institutions of incarceration. This requires making the structural changes that assure fair access to the justice system and fair treatment by this system and its agents. Just as the pervasiveness of the opioid crisis motivated the institution of treatment and counseling as alternatives to incarceration of opioid drug offenders, finding alternatives to locking up so many Black men is overdue and imperative.

We conclude with another reminder that many of the policies needed to correct inequalities between Blacks and Whites are broadly applicable to the needs and problems of other populations, too. Black–White inequalities are echoed in White–Hispanic and in middle-income White–lower-income White comparisons, too. Let's return to net worth as a useful and telling surrogate marker of many types of inequality. In 2016, White households had a median net worth that was about ten times greater than that of Black households and eight times that of Hispanic households. It is also notable that in 2016 the median household net worth of White households as a whole was about 7.5 times greater than that of lower-income White households (this segment includes about 25 percent of total White households). The fact that overall median net worth of Whites is four-fold that of the lowest-income quarter of the White population makes it clear that effective and need-guided policies for the elimination of economic inequality will be of a broad benefit and span a broad spectrum of the American population.

These considerations strengthen the argument that rather than being directed at particular racial or ethnic groups, these policies should address individuals according to need. This case is well and persuasively made by Sheryll Cashin in *Place, Not Race*. If new policies and appropriate corrections and modifications of existing social policies are instituted on the basis of need rather than race, they will benefit the appropriate and deserving members of all groups. They will certainly benefit some Blacks but also Hispanics, Native Americans, Asians, and a significant number of Whites, too. Beyond fairness, broadly applied, these strategies, inclusive of all in need and deserving, are politically wise and more likely to gain wide acceptance than those set up for the benefit of this or that specific racial group. Application of these policies would be highly constructive, remodeling the social fabric in ways that build better, more resilient, perhaps even happier, societies.

Suggested Readings
for *Thinking Race*

Alexander, Michelle. *The New Jim Crow* (revised ed.). New York: The New Press, 2012.

Although not intended as such, this book serves as a textbook for educating us about the nature and impact on communities of color of our system of mass incarceration.

Baldwin, James, and Margaret Mead. *A Rap on Race*. Philadelphia, PA: J. B. Lippincott & Co., 1971.

Two of twentieth-century America's leading social thinkers discuss race, each from a different perspective. This book provides an opportunity to listen in on a conversation between the era's leading Black writer and its most acclaimed anthropologist.

Cashin, Sherryl. *Place, Not Race*. Boston: Beacon Press, 2014.

Sherryl Cashin, a law professor and social commentator, provides a thoughtful examination of what should constitute fair college admissions policies, making a case for the replacement of racial and ethnic criteria with those defined in relation to need and disadvantage.

Christian, David. *Origin Story*. New York: Little, Brown and Company, 2018.

A leading exponent of "Big History" provides an engaging overview of what we know of the origin of everything, including the universe, humans, and their civilizations.

Coates, Ta-Nehisi. *We Were Eight Years in Power: An American Tragedy.* New York: Penguin Random House, 2017.

An edifying and satisfying retrospective of many of the themes that have made this thoughtful but pessimistic essayist one of America's leading commentators on race.

Du Bois, W. E. B. *The Souls of Black Folk.* Chicago: A. C. McClurg, 1903 (the Dover Thrift edition published by Dover Publications, Inc. is recommended).

A collection of essays on the position, plight, and prospects of Blacks as America moved out of the nineteenth and into the twentieth century. Now a classic, this work was an eloquent manifesto embracing assertion of equality over the accommodative postures championed by his rival, Booker T. Washington. It is foundational for the study of the sociology of Blacks in the United States.

Epstein, David. *The Sports Gene: Inside the Science of Extraordinary Athletic Performance.* New York: Penguin, 2013.

A review of the physiology responsible for outstanding performance in a variety of different sports, including track and field events, team sports such as baseball and football, and even competitive skiing. While acknowledging a role for genetics in athletic performance, this author is especially thoughtful and careful in reviewing the difficulties in pinning down exactly what genes might be involved.

Gates Jr., Henry Louis. *Stony the Road: Reconstruction, White Supremacy, and the Rise of Jim Crow.* New York: Penguin Press, 2019.

A dramatic and moving account and analysis of Reconstruction and its spawn, all foundational for the creation, shaping, and operation of the structural features of American society and economy that still operate to retard the progress of its Black citizens toward full equality.

Gardner, Howard. *Frames of Mind: The Theory of Multiple Intelligences.* New York: Basic Books, 1983.

Though contrary to the mainstream of current concepts of intelligence, Gardner's still-stimulating formulation of multiple intelligences is a reminder that there are many ways of being smart, not all of which can be measured by conventional IQ tests.

Gordon-Reed, Annette. *The Hemingses of Monticello: An American Family.* New York: W. W. Norton, 2008.

Thomas Jefferson and his descendants provide the family histories for a Pulitzer Prize–winning exploration of the complicated relationship of ancestry, culture, and race in the United States.

Hunt, Earl. *IQ and Human Intelligence*, 2nd ed. Cambridge: Cambridge University Press, 2011.
An excellent textbook on IQ that includes extensive discussion of the relationship between IQ and genetics and the variety of approaches and challenges encountered in its study.

Mann, Charles. *1491*. New York: Knopf, 2005.
This work describes the varied and vigorous peoples and civilizations that inhabited and flourished in the Americas prior to 1492 and the era of exploitation and colonization ushered in by the Columbian voyages. *1491* provides a highly readable correction of the notion that European expeditions arrived in the Americas and found a lightly populated wilderness free of civilization until its export from Europe.

Reich, David. *Who We Are and How We Got Here*. New York: Pantheon Books, 2018.
A highly readable, wide-ranging, and consistently surprising account of the wealth of revelations about human prehistory that have come from studies of DNA from bones of humans who lived thousands, even tens of thousands, of years ago.

Robinson Eugene. *Disintegration*. New York: Anchor Books, 2011.
An insightful description of the contemporary Black population as a collection of groups that differ in income, education, and prospects. Robinson explains how the experience of today's Black population differs from the Black community of the past that had a de facto leadership and its own newspapers and lived almost exclusively in socially and economically diverse Black neighborhoods.

Rothstein, Richard. *The Color of Law*. New York: Liverwright, 2017.
A compellingly written, insightful, and revelatory account of the complicity of federal agencies, laws, and policies that enabled, maintained, and even encouraged racially segregated housing in the United States.

Sandweiss, Martha A. *Passing Strange*. New York: Penguin Group, 2009.
A fascinating account of the reverse passing of a White patrician who assumed a Black identity, creating a Black family and episodically living with it while still periodically crossing back over the color line to resume his life as a prominent White man.

Shen Wu, Jean Yu-wen, and Thomas C. Chen, eds. *Asian American Studies Now: A Critical Reader*. New Brunswick, NJ: Rutgers University Press, 2010.
An excellent survey of a rapidly growing and increasingly influential group of American populations providing a helpful appreciation of some of the diversity that stands under the broad umbrella term of *Asian American*.

Shostak, Marjorie. *Nisa: The Life and Words of a !Kung Woman*. Cambridge, MA: Harvard University Press, 1981.
An anthropologist's vivid recounting of the life of an African Bushman woman as an example of being raised in a hunter-gatherer culture.

Stevenson, Bryan. *Just Mercy: A Story of Justice and Redemption*. London: Scribe, 2015.
A powerful, revelatory, and moving personal account by a man who looked at the injustice, brutal inhumanity, and pervasive racial bias of the criminal justice system and did something about it.

Takaki, Ronald. *A Different Mirror: A History of Multicultural America* (revised ed.). New York: Back Bay Books, 2008.
A carefully drawn portrait of the many different Americas that mix but do not always blend to make a multicultural America. Years ago, this author accurately predicted that the United States is only decades away from becoming a country in which we are all members of minority groups.

Treuer, David. *The Heartbeat of Wounded Knee*. New York: Penguin Random House, 2019.
A distinguished Native American author and academic provides a history of Native America. This history is often told through the personal narratives of American Indians, the central actors in this long drama, revising conceptions of them as living fossils, the remnants of a vanishing people. The idea of the vanishing Indian is replaced with themes of self-determination and sovereignty.

Vance, J. D. *Hillbilly Elegy*. New York: HarperCollins, 2016.
A highly readable book that provides a clear-eyed and telling look through the lens of family and personal experience into the life and psyche of some segments of working-class White Americans and their present situation.

Villaseñor, Victor *The Rain of Gold*. Houston, TX: Arte Público Press, 1991.

A novelist tells the story of three generations of a family, tracing the struggles, failures, and successes experienced by a family as it arrives in the United States, overcomes prejudice and exploitation, establishes the roots of a successful family tree, and succeeds.

Williams, Joan C. *White Working Class*. Brighton, MA: Harvard Business Review Press, 2017.

The author, a legal scholar and social commentator on work, gender, and class, provides an informative and revealing essay on the values, motivations, and concerns of what some see as one of America's most powerful ethnic minorities.

Index

143